HEAD CASE

HEAD CASE

How I Almost Lost My Mind
Trying to Understand My Brain

DENNIS CASS

HARPER PERENNIAL

NEW YORK • LONDON • TORONTO • SYDNEY • NEW DELHI • AUCKLAND

HARPER ● PERENNIAL

A hardcover edition of this book was published in 2007 by HarperCollins Publishers.

P.S.™ is a trademark of HarperCollins Publishers.

FIRST HARPER PERENNIAL EDITION PUBLISHED 2008.

Designed by Christine Weathersbee

The Library of Congress has catalogued the hardcover edition as follows:
Cass, Dennis.
    Head case: how I almost lost my mind trying to understand my brain /
    by Dennis Cass.
        p. cm.
    ISBN: 978-0-06-059472-5
    ISBN-10: 0-06-059472-1
    1. Brain—Popular works. 2. Evolutionary psychology—Popular works.
    I. Title.
    QP376.C37 2007
    612.8'2—dc22   2006043735

ISBN 978-0-06-059473-2 (pbk.)

08 09 10 11 12 DIX/RRD 10 9 8 7 6 5 4 3 2 1

For Liz

Either I'm dead right, or I'm crazy!

—Jefferson Smith, *Mr. Smith Goes to Washington*

The moment we arrived at his psychiatrist's office, my stepfather started acting sane. This was new. For the past twelve hours Bill had been cartoon crazy—howling like a wolf; lying on his back and bicycling his legs in the air; proclaiming, with equal parts terror and delight, "I can see the molecules in my hand." Transporting him from Queens to the Upper West Side had required physical coercion, but now that we were here Bill pretended like it was all his idea. When Dr. M. appeared in the doorway, my stepfather wriggled out of my grasp as if it were an insult.

"Dr. M.!" Bill said. Suddenly we were in nineteenth-century England. Bill shook his psychiatrist's hand as if we had traveled to the good doctor's country house by horse-drawn carriage. Dr. M. seemed puzzled by Bill's demeanor, but for me the loud voice, the theatrical formality, and the disproportionate enthusiasm made me doubt my reasons for bringing him here. I had known Bill since I was three, and in the past sixteen years this kind of behavior was well within the range of normal.

Dr. M. showed us in. Everything about his office signaled that this was where the right diagnoses were made—the messy intellectual energy of his paper-strewn desk, the rich, therapeutic smell of expensive leather. As for the doctor himself, he was squat and stolid and impressively bearded. As Dr. M. curled up

1

in his Eames lounger, I felt like you could search all of New York and not find a wiser animal.

On the phone, I had told Dr. M. that Bill needed to go to a mental institution. I felt I had no choice. In the past few years Bill had been in steep decline: trouble at work, trouble with the IRS, trouble with depression, trouble with drugs. I was home from what was supposed to be my sophomore year of college to help keep the family together, a job I was not capable of doing. Now my mom was alone in our apartment with my eleven-month-old half brother. We were both afraid of what might happen if I had to bring Bill back home.

As I detailed his transgressions of the night before, Bill sat next to me, his hands on his knees, his back erect. His body seemed to be engaged in the conversation, but his eyes were glassy and detached, as if I were telling a story about a man he'd never met.

Then Dr. M. did his intake. He asked Bill if he could remember how many Parnate he had taken the night before, or if he had eaten any foods that might have reacted with the Parnate, or if he had taken any other drugs, recreational or otherwise. Bill put on a show of taking each question very, very seriously before answering that as far as he knew there was nothing out of the ordinary about the past two days. Dr. M. then asked Bill how he felt about what I was saying, and, to my great relief, my stepfather revealed himself.

"We are both what you would call sentient beings, correct?" Bill said.

I looked at Dr. M. as if to say, *See?*

"And as such we are communicating with each other with this thing we call language," Bill said, sculpting his argument in the air with his hands. "Me speaking to you, then you to me, and through this shared exchange we arrive at an understanding of what is known, in layman's terms, as reality."

It would have been unprofessional for Dr. M. to look at me and circle a finger around his ear, but such a gesture would not have gone unappreciated. Instead Dr. M. remained neutral, and Bill—ever the sentient being—correctly read his doctor's passivity as a sign that his speech about the reality-constructing properties of language wasn't being well received. This made Bill intensify his hyperrational pantomime, but the harder Bill fought to appear sane, the crazier he sounded.

"This room, this office, if you will, is indeed a psychiatrist's office," he said. "And therefore it is reasonable to assume that a psychiatric determination is in the process of being made. I am cognizant of this fact."

Dr. M. had heard enough. "Your father has suffered a psychotic break," he said to me. Dr. M. said that the best thing for Bill was for him to receive institutional psychiatric care, but that there was a procedure to follow. First I had to take Bill to a hospital in Harlem for a physical. If his body checked out, then we would begin involuntary commitment proceedings. Bill nodded at all of this politely. This time, when he pretended to listen to a story being told about someone else, I suppose he was right.

"This is awesome," I said to Dr. M. "I mean, it's awful. It's really awful. But it's also awesome. Thank you, Doctor."

Bill and I left Dr. M.'s office and got into a cab. As we rode down West Seventy-ninth Street, Bill rested his head against the window. After fighting with me all night, he finally seemed to accept that he had lost. The lion had become a lamb.

And then he made a break for it. When we slowed to turn onto Amsterdam Avenue, Bill shouldered open the door, slipped out of the taxi, and started running down the street. The cabbie braked and I started to follow Bill out of the car, but stopped halfway out the door to marvel at the sight of him careening down the street. I had never seen anyone run like this. People on the street were staring, too. Bill was going fast, and hard, but there was something else, something extra in the running. Even if he had been wearing a tracksuit, there was no way this was exercise.

"Is he coming back?" asked the cabbie.

"I don't know," I said. "I don't think he is."

Underneath whatever practical concerns the cabdriver had about this fare, there was an implied question: What's wrong with him? He seemed to think I knew, but Dr. M. hadn't gotten into the why, and I had neglected to ask.

The next day I would get the medical explanation. Bill had overdosed on Parnate, which in large amounts can cause "agitation, mental confusion, and incoherence." From a clinical perspective this accounted for all of his symptoms. Bill had taken a double or triple dose, or possibly more. The rest was chemistry.

Still, this was one of those times when the simple explanation was also the least satisfying. The excess Parnate in his system might have triggered the psychotic break, but one overdose can't explain a life. Bill running down Amsterdam Avenue seemed like the culmination of years of slowly losing his mind. In 1980 he had moved us from Seattle to New York to take a job on Wall Street. He had tried to make a home in a city that was ill-suited for him. He abandoned friends and family and a decent life to pursue financial success, when by nature he didn't care about money. Maybe this madness was his fate and the Parnate had only provided the occasion.

Or maybe the answer was somewhere in between, a mixture of science and psychology, of biology and self. Thanks to Bill's

Parnate experiment, neuroscience had entered our house. Bill had been lecturing me on neurotransmitters and monoamine oxidase inhibitors and tricyclics, and reading books like Oliver Sacks's *The Man Who Mistook His Wife for a Hat*, and laughing at the neurological case studies of people who were farther gone than he thought he would ever be. Maybe the answer was all in his brain.

At that moment, however, still half in and half out of the cab, I wasn't thinking about explanations. I had one year of college, and with the exception of what I had picked up from movies, television, and Dickens novels, I knew nothing about how people worked, nor did I care. I wasn't even thinking in terms of pop psychology. As I watched Bill merge with the crowd and become part of the collage of windbreakers and shopping bags and haircuts, my thoughts were the product of a teenage mind that was best suited to reluctance and complaint. Why was I the one who had to chase him down? And what would happen if I let him go?

FALSE INSIGHT

I am not a scientist. When I was a boy, I went on field trips to the Seattle Science Center and watched the occasional nature show. I might have even enjoyed a book about dinosaurs. But by the age of twelve, whatever affinity I had for the scientific arts had turned into disinterest mixed with fear. Even though I went to a math-and-science high school, I hid in the English department, while in college I barely survived gut-level astronomy. Then science disappeared from my life for over a decade. Today the "latest findings" means an e-mail from a friend about the discovery of a 900-pound prehistoric guinea pig. Otherwise the sciences bring news that I would rather not hear—stories about deadly rays and faltering ecosystems and genetic betrayal. If science can't provide an easy laugh, then I do my best to avoid it.

If I weren't such a stranger to science, I doubt my idea to learn about my brain would have affected me so profoundly. It was the summer of 2002 and I was at my desk in my home office in Minneapolis, suffering from the worst case of writer's block I had ever experienced. All the office toys and charms that were intended to inspire—the picture my wife, Liz, took of my naked feet; my Greek good-luck eye; my Playmobil dragon—instead mocked me with their empty whimsy. My mental frustration was so powerful it manifested itself physically. I had a headache. My vision was blurry. My jaw hurt. Instead of writing, I passed the morning torturing myself with an internal monologue of self-rebuke. *I am truly astounded at how much you suck.*

Then my brain suddenly offered a simple, clear thought, a question that I heard in my head as clearly as if I had said it out loud:

How can you expect to live by your wits if you have no idea how your wits work?

This thought was accompanied by a positive emotional charge of a magnitude that I hadn't felt before, and haven't felt since. It was as if some kind of orgasmic *yes* juice were surging through every cell in my body. Centuries ago, the Greek mathematician Archimedes was taking a bath when he suddenly realized he could use water displacement to measure the volume and density of a solid object. He was so moved by his insight that he jumped out of the tub and ran naked down some ancient Greek street shouting, *"Eureka!"* I have found it! I knew exactly how he felt. After my insight I immediately left my office, went to the corner store, and bought an ice cream sandwich. For the next two days I was so pleased with myself that I did absolutely nothing.

Eventually this question about how my wits worked led me to the science of the human brain. At first my approach was casual. I started poking around the Internet, and what I found astounded me. While I had spent the nineties being angry at bands for selling out, neuroscientists seemed to have been making startling advances in our understanding of how the brain worked. I read about how brain scans might be able to detect potential terrorists, and about research into a neural prosthesis for storing memories, and efforts to make a monkey move a robot arm with his thoughts. Unbeknownst to me, I had been living in the middle of a cognitive revolution.

Once brain science was on my personal radar, it seemed like the entire world was taking part in a Mardi Gras of the human brain. There were brain-centric movies such as *Eternal Sunshine of the Spotless Mind* and *The Matrix* and *Memento*. Schoolkids by the thousands were taking part in Brain Awareness Week and writing poems with titles such as "Oh, My Sweet Hippocampus." I started seeing references to brain structures in places as disparate as *The New Yorker* and the supermarket tabloid *First for Women*. While shopping for hot sauce, I stumbled across a brand that featured a brain scan on the back label. That settled it. When condiments start offering signs and portents, I listen. The insight became a plan for living: I decided I would learn everything I possibly could about my brain.

But was this a good idea? Was the aha! moment a good indicator of a quality thought? Anecdotally it seemed that way. We owe the discovery of the ring structure of the chemical benzene, a prototype of the theory of evolution, and the theme of a symphony to insight. The fact that I had my insight in an office chair didn't discount its potential. The aforementioned insights could be credited, respectively, to sleeping, fighting a fever, and unwrapping a piece of cheese.

The science of insight, however, told a more cautious story. Insight sounds glamorous, ripe for a self-help book on how to think like a genius, but insight is only one of the cognitive tools the brain uses to conduct its daily business. "In real life, because you have an insight, it doesn't mean it's right," Mark Jung-Beeman, an insight researcher at Northwestern University's Brain Mapping Group, later told me. "You still have to go back and check your work."

If I had thought more critically about my insight, if I had kept a cooler, more scientific head, I might have answered my question the day I posed it to myself. The answer to "How can you live by your wits if you don't know how your wits work?" is this: It's easy. People do it every day. In the same way you don't need to know how a computer works to send an e-mail, or how an internal-combustion engine works to ride the bus, you don't need to know anything about the mechanics of your brain to live your life. The enthusiasm that can accompany a thought is meaningless: ideas are only as good as the outcomes they produce.

I knew this. I had seen what could happen when enthusiasm ran amok. I not only knew it, I had lived it.

One early summer day in 1979 my stepfather stopped me in the living room of our rented house in Seattle. He was sane back then, though still eccentric and erratic. Bill was an intimidating presence. Bill was a hippie, but he was not one of those sparkly, peace-loving pranksters whose eyes dance with love and mischief. Bill had black hair and severe eyebrows and cool blue eyes, and while he could be easygoing and playful, he also had a quick temper and a cruel sense of humor. When he first approached, I thought he was angry at me for not practicing the French horn. Instead it was very important to him that we talk. I needed to understand why he was considering interviewing for a job on Wall Street.

Bill's aspirations were relatively new. In 1977 he went back to school to get his M.B.A. Even at the age of nine, with my unsophisticated understanding of what suited a person's personality, abilities, and needs, this seemed to me an odd choice. When Bill married my mom in 1972 he wanted to be an actor, but he quickly abandoned that arduous and risky dream in favor of sleeping in, going on picnics, and generally letting my mom take care of him. One of Bill's favorite things was to get high and laugh at the poor quality of Japanese monster movies. Another was to listen to symphonies on his headphones and pretend he was the conductor. When Bill decided that he was going into finance, he may as well have announced—from the couch, a Tequila Sunrise in his hand—that he was going to become a professional tennis player.

Now Bill was about to graduate. He explained to me that there was a taxonomy of business school graduates. "I could get a job at a bank," he said, making it sound like McDonald's. "I could go work for Washington Mutual and be a VP someday. But that's for people who, you know, don't have what it takes." He let out a sigh. It apparently caused him some discomfort to talk about these lesser men, the kind who support their family and participate in their community and love their children.

"Now, Wall Street, hey!" he said. He was getting what he would call "jazzed up." He turned his fingers into drumsticks and played an imaginary high hat and snare. "New York, hey!" He snapped his fingers and clapped his hands. "C'mon! Get with it!" He then explained that if you were "hot shit," you went to New York. If you were "out of it," you went anywhere else. Bill said he was considering being a bond analyst—not, as he made it clear to me, a lowly stockbroker—at a place like Moody's or Standard & Poor's. Bond analysis was elegant and precise. Bill didn't use the word "poetry," but the implication was that he would be a kind of finance poet.

The rewards of this financial poetry, the "big bucks," would naturally make people take notice. Bill acted out a little play where he was coming back to Seattle after his New York triumph, and as he strolled down the street he tilted his head toward the people he knew, who in turn looked back with admiration and respect. More important, New York would let Bill be among people who were as cultured and refined and brilliant as he was always telling us he was. At long last, he would be appreciated.

Outside, through the picture window, I could see my friends from the neighborhood playing hide-and-seek. I wanted to join them, so I nodded and agreed. I didn't think this conversation meant anything. Bill was constantly talking. There was always another book or play or symphony for him to go on about, or he wanted to sing the new jingle for a fake product he had invented called Paralyzed Nun Pineapple Juice. The smartest strategy when he was on one of his rants was to hold very still; eventually both the topic and the man would go away. I never thought he would act on his words. Moving to New York seemed too far-fetched an idea to possibly come true.

I understand now that there were larger forces at work. Bill was thirty and feeling like now was the time to make something of his life. My mom was growing tired of his semiemployed lifestyle and wanted him to contribute more financially to the family. He was also being flattered by his professors, who were telling him that he had a special gift for business analysis, a finance poet's soul.

Ultimately, he fell in love with the idea of moving to New York because it made him feel smart. Now in my early thirties, I understood. My idea to study my brain also hit me at a vulnerable time. Liz and I were starting a family. With a baby on the way, it seemed that learning about my wits was important to more than my career. I had spent my twenties and early thirties becoming very accomplished at traveling, eating in restaurants, and talking about movies, but I didn't know if I had the skills to be a good dad.

I didn't feel like I had any glaring mental defects, but when I made an honest accounting of my brain, there were areas of, what I would call, concern. I thought science could help, even if the way I was thinking about science made me guilty of magical thinking. I approached my mental flaws like a man going to an herbalist in search of compounds and extracts that would, somehow, boost performance and well-being. I would address my worries about being scatterbrained and disorganized by examining the science of attention. In order to soothe my anxieties over becoming a new father, I would seek out the latest thinking on the neural mechanisms of fear and stress. As a chronic listmaker who had no confidence in his memory, I thought I could pick up some tips and tricks in that area of study. I made a loose logistical plan for my research, a shopping list for my mind. I was going to fully immerse myself in the world of neuroscience. I would visit labs, read scientific papers, and attend international brain conferences. I was going to be a human guinea pig whenever possible, trading my body for scientific knowledge. Then, in the end, I would put all these elements together and would know, with total certainty, exactly how my wits worked. I had only one rule: no prescription drugs.

At the time, I was mostly writing about popular culture, politics, and food, but taking on science didn't strike me as not doable. I never imagined that learning about brain science might be frustratingly difficult, or that the information I was about to absorb might undermine my view of the world, or of myself. I was so focused on the promise of the future that I didn't once consider that studying the brain might drag me back into my past. New York had ruined our family, but I had spent my twenties and early thirties putting all that behind me, even if the resolution wasn't entirely satisfying. My parents and I lived in separate cities (they were in San Francisco at the time) and we lived separate lives. As far as I was concerned, the apple had fallen so far from the tree it wasn't even an apple anymore.

My friends saw the risks better than I did. They knew that I was out of my depths, and that my project was not only a folly of my mind, but perhaps even dangerous to my soul. A number of them sent warnings that I might be making a mistake, but, buoyed by my insight, I ignored their concerns. I didn't know it at the time, but even though I wasn't running down the street raving, I was already starting to lose my mind.

"Are you sure you want to do this?" said my friend Clara. We were talking on the telephone.

"Of course I'm sure," I said. "This is a brilliant idea."

"I don't know," she said, her voice preparing for diplomacy. "It's going to bring up a lot of shit."

"Like what?" I said. A part of me knew exactly what she was talking about, but instead of listening, I made friendly, dismissive noises. "It's a story, Clara. I'm a *writer*. I'll just go out and learn about the stuff and then come back and write about it. What could possibly go wrong?"

MEETING MY BRAIN

I had spent my entire life without once thinking about my brain, but now that I was aware of it, I was still having trouble understanding what it meant to have one in my own head. Every day I read about how amazing the brain was—how there were more connections in the human brain than there were stars in our galaxy, how if you tried to duplicate the raw storage and processing power of a human brain using computer technology you would need many, many laptops—but I couldn't get my head around my own brain. The brain was like the Middle East—always in the news but hard to imagine—and before I started into my checklist of brain functions, I felt it was important to see one.

I went to the Minnesota State Fair. Known as "the great Minnesota get-together," each year the fair draws a million and a half people over twelve days. Fairgoers eat fried cheese curds and people-watch and soak up oddities such as the heads of dairy princesses sculpted out of enormous blocks of butter and portraits of celebrities made entirely out of crop seeds. On that hot August day in 2003, I was going for a serious purpose. No gorging myself at the Corn Roast and no laughing at teens in half-shirts prowling the midway. Today, after all, was "Brain Day."

As part of an effort to raise awareness about their neuroscience department and brain research activities, the University of Minnesota has put on Brain Day since 1997. The festivities took place in the University of Minnesota building, amid a souvenir stand selling Golden Gopher memorabilia and booths touting some of the university's other research programs. In the center of the room were tables and a busy staff. Each of the three tables had a whole human brain, a half-brain (with attached spinal cord), and jars holding animal brains, ranging from the rat (a very plump almond) to the sheep (a handful of oatmeal). One neuroscientist sported a whimsical baseball cap that looked like an exposed brain, while members of the university's press office handed out pencils with brain-shaped erasers to the cry of "Get your brain on a stick!"

Until now the brains I had seen were all drawings or 3D renderings. They were often accompanied by a lightbulb, as if to remind you that this was where thoughts came from, or dressed up with tennis shoes or eyeglasses. Cera Bellum, the mascot on the Brains Rule! website, has vampy eyelashes and thick, red, kissable lips.

The whole human brain in front of me was far less inviting. It had all the brain features you would expect, the gyri and sulci, the bulges and folds that give the organ its walnut-like appearance. But this brain had something extra: gore. The bottom was an open wound filled with the jagged ends of veins and arteries; the sulci seemed to be filled with black mold, and the whole thing was a sickly, pinkish gray, as if someone had stirred stomach medicine in a dirty ashtray. The university volunteers had neglected to anthropomorphize this brain, which would definitely have benefited from a little cowboy hat.

Erin Larson, a third-year graduate student in the neuroscience program at the University of Minnesota, presented me with a pair of latex gloves and asked me if I wanted to hold one.

"Not just yet," I said. The brain she tried to hand me was wet and shiny and reeked of formaldehyde. A piece of cellophane loosely clung to the cortex like peeling skin.

"I want to soak things in first," I said.

"Are you *sure*?" said Larson, as if the brain were a triple-fudge brownie. Larson handled these brains entirely unselfconsciously. She saw her first brain in high school, when a University of Minnesota–Duluth professor brought one to her psychology class. Now Larson worked in a lab that investigated whether or not gonadal hormones such as estrogen made female rats more vulnerable to drug abuse than males. She joked that she would either go into pure research or "sell out and work for a pharmaceutical company."

Larson had the kind of brain enthusiasm that I thought everyone was going to have. She thought brains were cool and she wanted other people to think they were cool, too, but she was largely alone in her excitement. The brains were more popular than the display promising new advances in mosquito control and the renewable-energy booth, but that wasn't saying much. Once in a while someone drifted over for a look, but a silent once-over or an "Is that real?" was usually enough to satisfy them. Up at the main demonstration table, the crowd seemed to be more interested in the brain-shaped erasers than the brains themselves.

To my surprise and disappointment, I wasn't reacting any

differently. That morning, in anticipation of seeing my first brain, I had thought to myself, *I am super pumped up.* I had come expecting a reprise of my eureka feeling. Seeing the brain would be like seeing the Grand Canyon, a feeling of awe at being in the presence of something magnificent. Instead I alternated between indifference and disbelief. I thought this was odd. Why would people's brains—including my own—not be naturally drawn to the brain?

At first I thought this was a matter of disgust, which, researchers theorize, plays a very specific role in human survival. When disgust researchers offer subjects foul-smelling stimuli such as sour milk or stale cauliflower water, part of the disgust response involves pausing to register the smell (the idea is that by recording the object of disgust, you will know to avoid it in the future). The way the spinal cord hung down from the half-brain like a ratty kite tail was certainly unpleasant, and the smell of formaldehyde seemed straight out of a disgust experiment. But neither of these sights or smells were bad enough to explain our mass reluctance.

Another possibility was "contamination fear." Subjects in contamination-fear studies are given "behavior avoidance tests," such as being asked to hold fake vomit, or a thoroughly clean but live earthworm, or being invited to eat miniature chocolate chip cookies served on an unused (and unarmed) rat trap. Contamination fear is sometimes linked to obsessive-compulsive disorder, which can manifest itself in intense worries about germs and viruses contaminating the environment. But I didn't think this was an OCD crowd, not with their cheesecake-on-a-stick and deep-fried candy bars. Given the level of consumption at the fair, I doubt there were many people here, myself included, who would have a problem eating a chocolate chip cookie off of just about anything.

If there was something repulsive about the brain it was more the idea of it. When a little blond girl in pigtails and with silver charms on her shoes approached the table, her mom seemed okay until Larson started spreading the word about how the brain works. Larson asked the little girl to stand on one leg, and when the girl complied Larson said, "That's your cerebellum doing that!" The little girl looked to her mom, unsure if this was an acceptable truth. Mom jerked the girl away and gave Larson and me a look like we were corrupting her child. "C'mon," she said to her daughter, "let's go get you a hot dog."

When the brain didn't offend, it turned people into uncomfortable comedians. ("That's what you look like on the inside, Paul!" "Oh, I don't think I have one of *those*.") Or people made fun of the animal brains for being small or for not having as wrinkly a cerebral cortex. "Look at that cat brain," snickered a guy who looked like he had been drinking since breakfast. "It's got one fold!"

One cultural anthropologist I talked to blamed this disconnect on organized religion. He said that in the seventeenth century, the Royal Society, the world's oldest scientific society, sold us all out. When its members wrote their charter, they purposely left out questions of the mind and spirit in order to avoid conflict with the Church. As a result, science has always shied away from exploring the physical side of the mind, while the Church has enforced its view that the soul was transcendent and immaterial.

Paul Bloom, a child development psychologist at Yale University and author of the book *Descartes' Baby*, has another theory. He says that we are all "commonsense dualists," meaning that the split between mind and body comes naturally to us. Bloom says people have two ways of thinking, one about the physical world, the other about the social world. Each way has its evolutionary advantages, but the two functions are separate, and perhaps irreconcilable. The social brain can't conceive of the physical brain, and the physical brain is invisible during social calculations.

"Our understanding is not that we *are* our bodies but that we *occupy* our bodies," he said to me during a phone interview. "We don't feel like objects."

Because we don't feel like objects, the brain—the object that does the feeling—doesn't seem real. The brain could be anything but *us*.

Maybe this was why some people had extra trouble connecting with the brain. After a while, Larson took a smoke break and I looked after her table. An intelligent-looking woman who casually passed by asked me what I had in the tray.

"A human brain," I said.

"Demon?" she said, puzzled but intrigued. Then I saw that she misheard me—"demon" for "human"—but it didn't matter. I couldn't see how anyone could think, even for a moment, that a Big Ten university would have a demon brain, much less put one on display, as if it were a sheep's or a kitty's.

"Demon?" I said. "I can't believe you said that."

She took offense. "How was I supposed to know?" she said.

I looked down at the brain in the dish. If only it would talk or twitch or crack a joke. *Mean something.* Instead it looked back up at me and offered nothing other than its own stinky ugliness. I realized that for all my talk I didn't really believe in the brain either.

"You're right," I said. "There is no way you could know."

Commonsense dualism aside, it shouldn't have surprised me that it was hard to reconcile mind and body. I grew up in a family that was all mind and no body. My mom never went to college but she was a passionate autodidact, devouring art, history, and literature as if she were a Ph.D. candidate. Bill was a self-styled intellectual and iconoclast. Even though he was a fierce atheist, Bill had his own religion, based on three tenets—zany, the Intellect, and being a Cass.

Zany was both an art form and a requirement for being of good character. Zany was dressing up in black, putting on sunlamp goggles and walking around in broad daylight, pretending to be from outer space. Zany was blowing up my stuffed animals with M80s on the Fourth of July. Zany was something as baroque as getting me to ask a department store clerk if they carried a line of menswear called "Death in Venice," or as simple as going behind a stranger's back, sticking out your front teeth, and making hillbilly noises.

The Intellect meant more than reading and thinking and educating oneself. Bill often used the word "possess" when it came to his Intellect, which by his estimation was both formidable and all-encompassing. To be "possessed" of an Intellect was to consume high culture and think critically, to be able to "grasp" almost anything, difficult concepts in particular.

But most important was "being a Cass." The Casses were small-town aristocracy. Bill grew up in Westville, Indiana. He was a baby boomer, the fourth of five children. His dad was president of the local bank, while his mother was a housewife who later became a schoolteacher. Bill grew up at the very top of the local food chain. Being a Cass described a code of honor, as well as high standards of diction and comportment. As a Cass,

Bill was, as he would say, "vastly superior" to other people, even when he was working as a clerk in a record store. His final judgment on someone else's poor taste or unseemly behavior was "A Cass would never do that."

I wasn't as confident in my Intellect, and I didn't care for zany, but I suppose it's possible some of Bill's grandiosity rubbed off on me, because sometime during Brain Day I became committed to the idea of communing with the physical brain, not just understanding it but *feeling* it, the way a spiritualist could not only speak with the dead but channel their suffering or their bliss. As the morning progressed, I started analyzing the demonstration brains to see if there was a clue in their physical properties that would help me with this side of my quest. Unfortunately, unlike other organs whose physical makeup suggests their purpose, the brain gives no indication as to its function. The heart looks like a pump, the lung like a bag for air, kidney meat looks like a filter. The only thing this brain looked like was a brain. It could be anything (paperweight, sponge), and it could be made of anything (ground-up breath mints, sugar cookie dough). In the second century, one of the brain's first anatomists, a Greek physician named Galen, thought the brain was made of sperm.

Maybe there was some clue in the brain's anatomy. Larson gave me the nickel tour, showing me the dangling brain stem that controls basic functions like breathing; the crinkly cerebellum that is crucial to motor function; the folded cortex that takes care of higher functions such as attention, language, and decision-making. Using the half-brain, she showed me the inner creases of what is colloquially called the limbic system, which plays a crucial role in emotion.

But the information didn't stick, in part because it all looked the same. The idea that stuck the most was that the more folded the cerebral cortex, the more thinking capacity it had. (The drunkard was right—you want a wrinkly human brain, as opposed to a smooth cat brain.) I ended up with a very rudimentary, and possibly grossly oversimplified, understanding of anatomy.

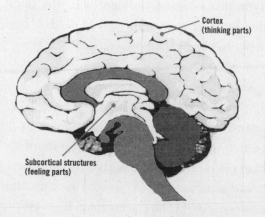

Cortex
(thinking parts)

Subcortical structures
(feeling parts)

Cross section of the human brain

Eventually I got over my squeamishness and held the whole human brain at Larson's table. The brain felt dense and heavy, but with a little give, like a Nerf football that had taken on water. Larson explained that this wasn't what a live brain would feel like. The formaldehyde "cooks" the brain, making the cells harder and stiffer than they normally would be. (Outside the skull, an unpreserved brain would spread like pudding.) Turning the brain over in my hands, hefting it, appraising it, I waited for a reaction. For a moment it was mildly interesting to think about how this was the engine that once drove a war hero or a soybean farmer or a high school principal, but mostly I felt more nothing.

Disappointed, I eventually left the building and did what I

normally liked to do at the fair. I ate two ears of roasted corn, some Tom Thumb minidonuts, an order of fried cheese curds, and a Pronto Pup. Feeling curious, I tried a deep-fried Snickers bar that was all the rage that year. Eating that Snickers bar will stay with me as one of my life's great regrets, but at least now I knew. Maybe that was the problem with the demonstration brains—they were too removed. Maybe I needed to see a picture of my own brain. Maybe then I would believe.

I started scouring the volunteer boards at the university and local hospitals. Often the flyers featured a clip-art brain spiced with a festive bolt of electricity, but usually it was all business. The postings for these experiments looked exactly like the coffee-shop ads for roommates and bass players, except in this case the insomniacs, depressives, and methamphetamine addicts who answered were presumably put to better use. *Looking for smokers with no history of mental illness for imaging study. Must be right-handed. Pay is $60 for two hours in the scanner. Call Marsha,* and at the bottom, a crispy fringe of tear-away tabs with a phone number.

Standing at these boards, I often felt like I was trying on problems. I'd gotten pretty annoyed at Jiffy Lube the other day—did that qualify me for a study about fits of uncontrollable rage? I hadn't done a very good job keeping in touch with my best friend from high school—did I have problems forming lasting personal relationships? There were studies for drug abuse and memory and low vision, and I collected the ones that made reference to imaging work and started calling. I probably wasn't sick enough to be a subject, but surely my brain would be in high demand as a healthy control.

Getting in proved harder than I thought. Whatever social skills you have are worthless. Researchers don't care about you; they only care about your brain, and it turns out they have very discriminating tastes. Any history of epilepsy? they asked. How about cerebral palsy, stroke, cancer, lazy eye, encephalitis, manic depression, or eating disorders? Have you ever had a learning disability or been in speech therapy? Have you ever taken anti-depressants, or thought you were drinking too much? Halfway through the screening calls I found myself thinking, *Picky, picky.*

I would be doing fine until I got caught on something like the speech therapy to get rid of my lisp in fifth grade or the hand-ful of months in college I'd spent on antidepressants and they shut the door. I once wasted two hours taking a Structured Clinical Interview for DSM-IV (SCID), a Heinrich-Carpenter Quality of Life Scale, a Premorbid Adjustment Scale, a Family Interview for Genetic Studies, and a Mini Mental Status Exam only to be rejected. The lab sent me a letter that said I was "well groomed and dressed appropriately" and "displayed appropriate affect and made good eye contact"—how could they not want me? I tried to bargain with them, but my brain's viability, unlike my self's, was not something I could haggle over. I was either within acceptable tolerances or I was not. When yet another lab assistant rejected my brain for an Ecstasy study, I asked, "Do you want someone who has taken fewer drugs . . . ?" and she cheer-fully replied, "Or more!"

Eventually I got into to Stephane Lehericy's lab. On a crisp fall morning I paid a visit to the Center for Magnetic Resonance Research at the University of Minnesota. The Center for Mag-netic Resonance Research is a low-slung building on the edge of campus. When I pulled into the parking lot, the first thing I no-ticed were the grain silos in the background. The second thing I noticed were all the warning signs. Apparently a high-powered magnet can do things to the human body you would normally associate with knives and guns.

Inside, it was plush. Whereas most of the science buildings on campus are dull linoleum-and-cinderblock affairs, the CMRR is *designed*. It has a proper lobby done up in glass and wood. Everything about it announces that this isn't a place where scientists make findings, they make *advances*.

Lehericy is a handsome man from France. In the movie he will be played by Richard Gere. I signed forms and put my wallet and keys in a locker so my credit cards wouldn't become demagnetized and so there would be nothing in my pants to fly across the room toward the scanner and cut someone to pieces.

Lehericy then led me downstairs to room 100-12, which was divided into two parts. Both sections were white and cool. In one half there was a control room with a dry-erase board, an observation window, and banks of computer equipment. In the other room there was a Siemens 3-Tesla Trio magnetic resonance imager, an eight-foot-high white Life Saver that hummed and knocked like an old car.

Lehericy explained his work to me. He said he was researching the role of the basal ganglia—a small cluster of neurons deep in the brain—during motor function as it might apply to Parkinson's disease. I had to hand it to him: he didn't dumb it down.

This was my first real experience talking to a scientist on his turf and it pointed to a problem in understanding the brain that went beyond commonsense dualism. Any time you are talking about the brain, you have to first establish which brain you are talking about—genes, neurons, clusters of neurons, populations of neurons, the whole brain? Or are you talking about the parts that aren't even neurons, the neurotransmitters and enzymes and glia? The study of the brain is like the fable of the blind men and the elephant, only the blind men number in the thousands and the elephant is a three-pound pile of quivering goo.

I gaped and stared and pretended to understand and then finally admitted to Lehericy that I didn't know what he was talking about. In an effort to help, he drew an impromptu brain on the dry-erase board. He then went off, drawing arrows and squiggles and scrawls. This was supposed to help illuminate his point, but I was more fixated on how much the brain he drew resembled a cinnamon roll.

Then it was time to be trained. "Do you play piano?" asked Lehericy. He was holding a keypad in his hand with four square buttons on it. Was this a final qualifier? Were the piano lessons I took my senior year of college going to disqualify me because they were looking for musically naïve subjects?

"I've been in some bands," I said. "Mostly I played bass. But a little guitar. And some keyboards."

Lehericy didn't know what to do with this, either because he was from France or because it was an irrelevant thing to say. What he was less thrilled about was that I am left-handed. For the sake of consistency, scientists generally only study people who are right-handed. Nevertheless, he trained me in. The task was very simple. I would hear a series of tones and each time I heard a tone I would tap my finger in the following pattern:

Index, ring, middle, pinky, ring, index, pinky, middle, ring, pinky. Tap, tap, tap, tap . . .

Once I had the hang of it, Lehericy put me on the scanner bed. Up close the scanner looked less like a single Life Saver and more like the whole roll. I lay down, and Lehericy put a wedge-shaped pillow under my knees and packed my head with a vacuum bag filled with tiny Styrofoam balls so I couldn't move.

"Is the magnet on?" I asked.

"The magnet is always on," he said, rolling me inside. I now knew what it was like to be put away in a drawer.

Compared to a typical hospital MRI, the CMRR houses complicated experimental machines, but the basic principles are the same. An MRI takes static pictures, but a functional MRI (or fMRI) like the one I was in adds the component of time by tracking blood flow. The thinking goes that where there is blood there is energy, and where there is energy there is meaning.

Now that it was too late, I took a moment to wonder if putting my brain inside a high-field, experimental magnet was a good idea. The night before Liz had joked about them finding a tumor, then took it back. We were pregnant now. We were signed up for baby classes and everything. I was going to be a dad and here I was subjecting my brain to who knows what kinds of abuse. Lehericy had assured me that the scan was safe, but then again I had also signed all kinds of forms stating that I wouldn't be mad if something went wrong. What was I doing here?

As it turned out, the experience was anticlimactic. The task was even more boring in the scanner than it had been during the training. "Go," said a recorded voice with a French accent. And I went. Index, ring, middle, pinky, ring, middle, index. Tap, tap, tap, tap . . . Index, ring, middle, pinky, ring, middle, index. Performing was harder than I thought, mostly because the experiment was too simple. Despite the noise, a lot of people fall asleep in fMRI machines, and it's not because of the magnet but the mighty force of boredom. The brain needs more than tones to keep itself awake.

I was in the scanner for around two hours. I lost count of how many trials, and I think I drifted off for a while. It felt like a waste of time. Here I was in a multimillion-dollar experimental imaging machine and all I was doing was tapping my fingers. Even if I wasn't a Cass, it seemed that my brain had more to offer than this.

After the session, Lehericy showed me my anatomical scan. The functional images would take a few weeks to compile. We stood in front of a computer screen that showed two views of my brain, one from the left side, the other from the top down. Both images were white and luminous, and I could see the individual gyri and sulci, as well as the anatomical structure of my spine.

"So I have a brain," I said.

"Yes," said Lehericy, "you have a brain."

Using a mouse, Lehericy dragged a cursor across the image, changing the view from the back of my head to the front, from bottom to top. The level of detail was remarkable, and there were moments when I felt like I was tunneling through my brain's cracks and fissures, rolling over its different nuclei. I had recently seen an ultrasound of my son, and even though he looked like little more than a lima bean in a charcoal cave, seeing him made him more real. But this brain—my brain—was it real? I felt like I was back at the state fair. Whether it was commonsense dualism or disgust or some Cass-like reluctance to being reduced to a physical being, I didn't believe this brain was mine. I found this disturbing. Even though not feeling your brain is a perfectly healthy and normal thing, I thought that there was something sinister in how my brain denied its own existence.

When a feeling of recognition finally came, it wasn't of my brain but my face. In addition to showing the intimacies of my brain, the images in front of me also captured my skull and neck and spine and teeth. It was, after all, an image of my whole head; we just happened to be focusing on the insides. But around the edges of the scan were shapes that were more recognizably me, and each time the cursor moved, my profile morphed and warped. Now my nose was bursting, then it disappeared. My teeth and tongue faded in and out; my mouth puffed and thinned.

"Do you have any questions?" asked Lehericy.

This was my chance, but I had nothing. I looked at the wrinkles and dimples of my cerebral cortex.

"Do I have good folds?" I said.

"You have a nice healthy brain," said Lehericy. Maybe it was his French accent, or maybe it was my own fledgling paranoia, but the way he said it sounded an awful lot like a question.

THE PREHISTORIC BRAIN

I took this inability to commune with my brain very personally. Even though Lehericy assured me I had a healthy brain, the seed of doubt had been planted. I had started my research believing the human brain was a wonderful thing, prone to weakness at times, but mostly capable and nimble and strong. Overall the design was a good one, and I assumed that the organ was fresh and modern and up-to-the-minute and highly evolved. Maybe this was a mistake. Maybe the human brain was a piece of junk.

When I took a big step back, it seemed my suspicions were confirmed. The brain was not as shiny and new and purposeful as I had hoped. In fact, the brain was largely an accident. Recent research has found that roughly two million years ago, thanks to a genetic flaw, a "disabling alteration" on gene MYH16, our ancestors were born with smaller jaw muscles. This diminished ability to chew led to a smaller jaw, which led to more room in the head for a larger brain. There is some debate as to when the human brain completed its expansion—or even if it has stopped changing—but the gist is that while cultures and civilizations have evolved and changed, the physical brain has not. In the words of evolutionary psychologists Leda Cosmides and John

Tooby, authors of *Evolutionary Psychology: A Primer*, "our modern skulls house a Stone Age mind."

Stone Age doesn't mean simple. Write Cosmides and Tooby, "In saying our modern skulls house a Stone Age mind, we do not mean to imply that our minds are unsophisticated. Quite the contrary: they are very sophisticated computers, whose circuits are elegantly designed to solve the kinds of problems our ancestors routinely faced." But I took it to mean that even though I lived in the twenty-first century, my brain was operating as if it were the dawn of time. The progress of the human experience was not a function of biological advances in our brain. Everything around me—from the Constitution of the United States, to the supercomputer, to the plastic nose at your local convenience store that dispenses candy-like snot—were the products of brains that hadn't changed in tens of thousands of years. I found this a little insulting, not to mention disturbing. This seemed like a long time to go without an upgrade.

Evolutionary psychology is an attempt to figure out how these prehistoric systems work, and then apply that knowledge to how people think and behave. Apparently my brain was designed to solve hunter-gatherer problems: Where do I get food? How do I avoid suffering and death? Who would be cool to hang out with? The context in which it solved those problems—i.e., this crazy, mixed-up world—was window dressing.

Often, evolutionary psychology focuses on reproductive success. Satoshi Kanazawa at the London School of Economics, for example, recently published a paper called "Why Productivity Fades with Age: The Crime-Genius Connection." Kanazawa notes that statistically both scientists and criminals are most active in early adulthood. By the time they reach forty, productivity drops off, caused, according to Kanazawa, by marriage and family. Once there is no longer a reproductive need, the drive for success wilts. The exceptions, in the case of the scientists, are those who stay single. These men are more professionally active into their forties, as are the criminals, or at least the ones who stay out of jail.

Evolutionary psychology is not without its critics, who find it circular and reductionist, with theories that are impossible to verify. The late, great Stephen Jay Gould considered it cocktail-party science, but what did he know? I was in love with this theory—this was science for English majors.

In light of evolutionary psychology, Bill's decision to take us to New York certainly looked different. What seemed insane— the sudden change from zany hippie to success-driven yuppie— was quite sane. Bill was not taking part in some embarrassing Zeitgeist, but was instead in the throes of a reproductive panic. He was almost thirty. Whether he realized it or not, if he was going to have his own child, if he was going to pass along his own genes and not merely play custodian to some inferior step-genes, then his prehistoric brain was going to have to get off its ass and make a play.

If Bill couldn't evolve, he could adapt. He could make an effort. My parents still went to parties, but they started appreciating wine instead of downing Tequila Sunrises. They quit smoking, both cigarettes and pot. Bill ran three times a week and completed his homework assignments on time. My mom made large, fresh salads. We ate them.

"Just imagine," Bill said. "Lincoln Center, Carnegie Hall, the Met! A real opera company and a real symphony orchestra!"

We were in the kitchen, having dinner. His graduation was a month away, and the talk was of what to do next. My mom had made coq au vin and a salad, and they were sharing a bottle of wine. Hanging over the breakfast nook was a framed poster of the Seattle Opera Company's production of Wagner's *Ring* cycle. When Bill talked about New York's orchestra, he made conducting gestures in the air. Then he stopped to butter a slice of Roman Meal bread.

Biologists talk about "optimal foraging theory." Under optimal foraging theory, an organism reads the environment and feeds accordingly by taking into account how different foraging sites will produce different results under different conditions. One evolutionary biologist notes that "a well-designed mechanism should be sensitive to these differences in rates, detect changes in them, and take into account the statistical uncertainties inherent in a limited number of observations." Which is a complicated way of saying that even a dumb animal knows what's good for it.

I don't think Bill was being very sensitive to differences in rates or accounting for statistical uncertainties. For Bill, New York was relentlessly positive, the most optimal place to forage he could possibly find. In conversation he switched between historic New York and literary New York and the New York of black-and-white movies. One minute he talked about Henry James, then about how Vaux & Olmsted created Central Park, then a scene from the old movies he made me watch, ones where women wore pantsuits and smoked and talked fast at their newspaper job. Everything except what it actually might be like to live there.

My mom wasn't saying much, but she went along. She pulled her shoulders up to her ears in sympathetic excitement. I couldn't imagine the New York Bill was describing. New York in the movies I liked was a dark, nighttime place that consisted almost entirely of dangerous public parks. I had read one book set in New

York, a story about a mouse and a cat who lived in the sewer under Times Square. They were always cold and hungry.

"New York is so far away," I said. This was directed to my mom.

"That's the best part," she said. "People can visit."

"I don't want to move," I said. "I like Seattle."

Bill made it clear that I was wrong about Seattle. We had talked about that already. Seattle was small, ignorant, and provincial. My mom's shoulders were up by her ears again. Then she used our family slang for exciting excellence.

"It'll be spicy," she said.

If I were a little older I might have figured out what was going on that night. Regardless of whether it was a good decision or a bad one, whether made for reasons of arrogance or greed or some hidden evolutionary drive, they had already decided. This discussion was largely for show. We were, in effect, already in New York.

"A true world city," said Bill, as if he were at that very moment walking down the gangplank of a steamship in New York Harbor. "Like London. Or Paris. Or Rome. The architecture alone. The artists and intellectuals. My God, can you imagine it? People who actually *read* and *think*."

Of course, New York wasn't exactly like this, but even if Tom Wolfe came to our house and told Bill personally, I doubt he would have listened. Nor could he know that even thinking of New York in this way was misguided, or that his drive to the city was based in something old and deep in his prehistoric brain. (As Cosmides and Tooby write, "We all suffer from instinct blindness.") If there was some basic brain module that protected its owner from overreaching, Bill didn't have it. But I did. I knew that there was something dangerous in how Bill was thinking about New York. Nothing that sounded that relentlessly positive could come to any good.

If I was uneasy about having a prehistoric brain, I dealt with it by turning this knowledge on other people. One of the chief joys of evolutionary psychology was that it was easy and fun to play with; even more so than that old favorite, pop psychology. Evolutionary psychologists sometimes talk about the brain having "modules," and this seemed like a good way of looking at other people's behaviors. You pick someone in a crowd and imagine them wearing a saber-toothed-tiger pelt and carrying a sharp stick and then imagine whether or not they would have thrived.

There was the temptation to turn evolutionary psychology on those who were closest to me, but I knew better. It was October of 2003, and our baby was due in less than two months. Liz had been buying baby books in bulk, and we had signed up for three baby classes, including the warmhearted and delightful All About Babies (featuring a video unexpectedly hosted by actor/director/mensch Rob Reiner) and the stressful and incapacitating Infant CPR. Our first class was called Childbirth Preparation, which struck me as slightly clinical and grossly optimistic. Was such a thing even possible?

On a Saturday morning we joined a small audience gathered in an auditorium in Abbott Northwestern Hospital. There were around forty of us, twenty couples and one very uncomfortable single mom. We were all hoping to become baby-wise and learn how to baby-whisper. All of us, except for Travis.

Travis's prehistoric brain simply could not keep its owner's body still. Everything—the sound of a soda bottle opening, a flicker in the overheads—caused Travis's body to jitter and twist. His wife, who seemed pregnant enough to deliver at any minute, tried placing a calming hand on Travis's thigh, but even her gentle touch only set off another spasm, which expressed itself in the rustling of Travis's black University of Wisconsin nylon wind pants. Those pants were quite eloquent that morning. They spoke volumes. Travis's every movement was broadcast in nylon, which whistled and zipped and zipped and whistled. It was like sharing a room with a bird.

The speaker, a middle-aged nurse with a firmly informative tone, announced that we were now going to watch a video. Travis let loose a throaty "Sweet!" then frowned when he learned the video would contain scenes of a live birth. When the nurse explained this was going to be one of many such scenes this morning, Travis's already scrunched-up face commenced to eat itself. Then he brightened. His prehistoric brain had an idea. Travis took out his Palm Pilot, rolled his shoulders forward, and did the sensible thing. He played Tetris.

When Travis's wife discovered this, she was not pleased. She turned to him and started in, quietly, gently. I couldn't see her face—only a wall of shoulder-length dark hair—but I could imagine the tone and content. With her pregnancy, if it was anything like ours, had arrived a sense of *great urgency*. This need to put one's life in order before having a baby is called the "nesting syndrome," but it is not always a cozy phenomenon. What it boils down to is that there is absolutely zero room—zero room, do you hear me?—for bullshit of any kind. A husband's personality tics that at one time might have even seemed charming now simply did not stand.

"What's the point?" Travis said when his wife tried to get

him to come around. "It's gonna happen anyway." He was refer-ring, I assumed, to the Miracle of Life.

His wife started in again. Travis was about to protest, but she soon got through to him.

"Oh," he said. "It makes you feel better to learn. . . . Oh. . . . Okay."

Knucklehead, I thought, then I turned to Liz to see if she was laughing, too. Instead her disapproving smile was directed at me.

"Are you taking notes on the class or on *him*?" she said.

I had options, ranging from coy to disingenuous to outright untruthful, but instead I made a face that strongly suggested fu-ture good works and turned my attention back where it be-longed. As much as I wanted to distance myself from Travis by making fun of him, I was far more like him than I would care to admit. True, I had a better haircut. And I did not look like a turtle. Nor would I ever take a birthing ball and dribble it in front of my wife as if challenging her to a game of one-on-one. But these were minor differences. I had more than my share of neurologi-cal flaws—or at least flaws that could be described in neurologi-cal terms—and rather than turning the searing Eye of Science on poor Travis, I reminded myself that the point of this project was to help me face my own all too human brain.

It seemed to me that this little episode with Travis called into question the limitations of the brain. If we couldn't pay atten-tion to a video about babies—which, given the circumstances, seemed like some optimal foraging—then how were we going to manage the real thing? And what about the compounding of small moments like these? The more I walked around with the idea of my brain being tragically old, the more it seemed that human beings couldn't even begin to solve the problems we've created for ourselves.

I used to think the human mind was capable of understanding anything. Then I visited Janet Metcalfe's lab in the psychology department at Columbia University. One of Metcalfe's interests involves metacognition, which is how people think about how they think. While seated at a computer in Metcalfe's warren-like lab, I was given Spanish vocabulary words to study. A Spanish word would pop up on the screen accompanied by its English translation. I had as much time as I wanted to study it. Then, a quiz. After the quiz I was given the words I got wrong and asked to choose which ones I thought I needed to study again for the retest.

I took Spanish in high school, so many of the words were readily studiable. *Librocambio* (free trade). *Picaporte* (door latch). But some of these words. Wow. Like *tejemaneje* (bustle) or *gualdrapear* (to flap) or *zangarriana* (migraine). They would fly out of my mind the minute I saw them. Bridgid Finn, one of Metcalfe's graduate students, told me later that some of the harder words were from obscure Aztec-influenced dialects that were only spoken in a handful of villages in Mexico. She had a hard time with them herself, and she was fluent in Spanish.

During the first trial I tried to study all the words, regardless of difficulty, but after a while I said *Fuck it*. By trial three I stopped even looking at the hard ones at all, lest they undermine my grasp on the easy and medium easy ones. Not that this felt good. Even though my data would be folded in with other data and averaged out or run through logarithmic analysis protocols or whatever, they were still collecting information about my performance. I was still being watched, judged, and because an experiment's task often stretches the limits of your brain, I found myself experiencing what I call experiment shame.

Afterward, I went to Janet Metcalfe's office for the official postmortem. It was a nice office with high ceilings and a view of Columbia's grounds, which included an original casting of Rodin's *The Thinker*. On her white board there were big squiggly shapes, her four-year-old daughter's "Sad People" characters.

"When I found out the test wasn't going to be multiple choice, my heart dropped," I said.

"It's very hard," she said.

"I knew I could recognize *pieces* of words, but . . ."

Metcalfe asked me which words I chose to study, and I felt like apologizing for easy choices like *botella*, for bottle, or *calcio*, for calcium. But once I verbalized my decision I instantly saw the point of the experiment. It wasn't a juicy aha! but an insight nevertheless. Our prehistoric brain's ability to learn has limits, and those limits are largely self-determined. You pick, often unconsciously, what you're comfortable learning. What is hard and what is easy is not an external property of what you're studying, but an internal dialogue with your own mind.

"What you are saying is exactly what we've been finding," said Metcalfe. "People choose to study things they think they are going to benefit from. And we talk about that as a region of proximal learning."

This notion of proximal learning seems obvious when it's pointed out, but the actual mechanism works very subtly. The region of proximal learning does not include facts or ideas that you already understand—I didn't study the handful of Spanish words on the list that I already knew—but begins with the tender edge of the unknown. According to Metcalfe, the metacognitive challenge is trying to figure out what you've mastered and what remains a mystery.

"According to our research, what you should do is push from your level of expertise," Metcalfe said. "Take your strengths and push from them, and push and push and push. You shouldn't jump in and expect you're going to be able to just get it. If you do, you're setting yourself up for what in the literature is called 'labor in vain.'"

Our first night of laboring in vain in New York was New Year's Day, 1980. Bill wanted to give me a quarter of a Valium because I couldn't seem to make up my mind between laughing and crying. My mom resisted. The debate that followed was less over whether or not the Valium should be dispensed and more about in what amount. Bill said a whole. My mom said a quarter. The discussion was rational, almost academic. When your eleven-year-old is freaking out due to a cross-country move and you decide to give him benzodiazepines, what exactly is a responsible dosage?

We were in a motel near Newark airport while our belongings, including our 1974 orange VW Bug convertible, were being trucked across the country courtesy of Standard & Poor's. We had two days before the moving truck was set to arrive, and we had nothing but a television and one another. My parents had already sedated themselves, and they reasoned this was the best course of action for the whole family.

After I got my Valium, we turned our attention back to the TV. Bill was taking special delight in zany New York television commercials. There were people singing a showtune about some place called the Milford Plaza, commercials for an ice cream cake, called Fudgie the Whale, and designer jeans ads with new wave sound tracks and asymmetric story lines. Had we landed in Europe? Had we traveled to the future? Why was everyone singing and shouting? Everything seemed significant and yet nothing was meaningful. Bill thought this was delightful. I read the messages from the TV as warnings that we didn't belong here, but then the Valium started to take effect. A quarter of a tablet did quite nicely.

Two days later we drove to our new home. Haworth, New Jersey, was a small town, but it seemed outlandishly wealthy. In Seattle, small towns meant Eastern Washington, spread-out places populated by people who inspired Bill to make galoot noises. When we pulled up to our house I thought Bill had rented us a castle. It was a Tudor with a built-in garage and a yard that was so big it felt like we had grounds. Inside it was even more impressive, in part because it had two stories. No one we knew other than my dad's well-off sister in St. Louis lived in a place that had a second story, but we now had two floors, and the living room ceiling went all the way up to the top of the house. Not only was there a staircase—my first—but it was grand and exposed as it ran up to the second floor. There was even a small balcony overlooking the living room in case you needed to make a speech.

At first the alien territory was exciting. Every day Bill brought home stories about the great costume drama that is New York. A three-card monte game gone awry, or the temerity of drug dealers in a park near his office, how they called out "Smoke, smoke, smoke," right there in broad daylight. Or he would talk about the Hasidim in the Diamond District or the capos in Little Italy or even something tiny, like how a coffee "regular" meant not black as you might think, but with cream and sugar already added in.

"Look at this," Bill said one day when he got home. He handed me a Drake's Ring Ding.

"Weird," I said. "It's like a Hostess Ding Dong but it's called a Ring Ding."

"I know," he said, laughing. "Can you believe it? A Ring Ding! Absurd!"

This was New York. Not the museums or the restaurants or the symphony. He had come for the glory of New York and ended up focusing on a snack cake.

Once Bill took me to work, but instead of showing me around the office, he wanted to show me Wall Street, as in the street itself. He took me around the narrow, winding streets and explained why they were so narrow and winding. There was an old church, dark as burnt toast, that I was supposed to pay attention to, but not the people. Everyone around us was invisible. Other human beings were not in Bill's region of proximal learning.

This was causing problems at work, where Bill's technical abilities were beyond reproach but his office politics were dismal. Bill never put it this way—that would be admitting some kind of flaw—but the social defeats at S&P were affecting him. For one, his coworkers were eating into the fantasy of a New York populated entirely by writers and intellectuals. The boors weren't just at the bus stops and at the pretzel stands but in the halls of S&P, which for some odd reason he thought was going to have more of a salon-type atmosphere. He didn't understand why when he spent time at work drawing a cartoon of Brother Roberto, a Franciscan monk who was also secretly an insect, his coworkers didn't applaud his genius.

"It's like they don't get me," he said once, not thinking that maybe he didn't get them.

If only there was some kind of trade-off between work and play, but other than marveling at Drake's Cakes and the occasional piece of street theater, my mom and Bill didn't do much. They were engaged in suboptimal foraging. When I pointed out that neither of them seemed to be getting much out of the city, Bill took great offense. He defended the decision to move, stumping for New York from our living room in New Jersey. "Seattle was so out of it," he said. "They didn't have a world-class symphony orchestra, there was no *theater*, no *art* . . ." He looked for something else, some kind of capper. "What about the Cloisters?" Then he calmed down and we watched network television.

We were miserable. Bill started getting hives from the stress, then both of them started to get migraines. There were hurried trips to the emergency room. Sometimes both my mom and my dad would have migraines at the same time, and they would argue over who was responsible for driving to the doctor. But still they hung on to the idea that this was where they were supposed to be. My prehistoric brain was crying, *Mistake, mistake*, but Bill's brain blamed it on the Paramus Park Mall. "We never should have moved here," he said, referring to New Jersey, not to the East Coast. "It's so out of it." Bill was back to selling us on the city, only this time he had refined his pitch. He said we needed to be as close to Manhattan as we could afford. So after one year in New Jersey, Bill solved all our problems by moving us to Queens.

INATTENTION TO DETAIL

In late 2003, with the baby coming soon, Liz and I were receiving dire warnings from people who were already parents. The most specific piece of advice we got was to see as many movies as we could right now because we were about to fall off the face of the earth. Otherwise the news from the other side ranged from the vague ("It's hard") to the vaguely ominous ("Everything is going to change").

If having a baby was anything like learning about your brain, then I was in trouble. My experience with Lehericy was repeating itself, as every researcher I talked to left me feeling like I was actually losing knowledge. I got yelled at by a gambling-addiction researcher when I didn't understand that compulsive shopping and compulsive gambling activated the same basic reward networks in the brain. "You have to think fast," he said, chopping his hands at me. Another source turned out to be a hard-line behaviorist who dismissed the entire premise of my question about my wits with a thundering, "I don't believe in the brain!" At the 2003 Society for Neuroscience's annual meeting in New Orleans I stumbled around in an information-induced haze. The first night I passed out—in my clothes, at 7:00 p.m.—after doing nothing more than *talking* to people about science.

The hardest part of my adventure into neuroscience seemed to be simply paying attention. I loved chatting with neuroscientists—they were often fascinating people—but I couldn't last more than five minutes talking about the brain before changing the subject to music or food. Rather than chalk it up to being outside my region of proximal learning, I started wondering if there might be something wrong with my attention network.

I called Barry Giesbrecht, an attention researcher and assistant professor of psychology at the University of California–Santa Barbara. What he told me spoke directly to my concerns. Giesbrecht said the attention system could be characterized by two central dichotomies: reflexive versus voluntary, and focused versus diffuse.

"Think of it like a tightrope," he said. "When you're sitting at your keyboard, your attention is highly focused on the task. The benefit is . . . all your resources. The cost is . . . being less aware. On the other side of this tightrope is a state of diffuse attention where you're trying to attend to everything in the environment. You can be aware of gist, but you cannot be completely aware of all the minutiae."

I said that the problem with attention, then, is that we are often diffuse when we need to be focused and focused when we need to be diffuse.

"That," said Giesbrecht, "is a failure of cognitive control."

Unfortunately, failure of cognitive control wasn't something I could blame on my parents or my upbringing. My mom was an avid reader and always did well at her job. Office-politics problems aside, Bill attacked his work with a ferocity and skill that made up for a lot of Brother Roberto cartoons, and when it came to feeding the Intellect, there was no limit to his concentration.

After we moved to Queens, Bill began to devour the city. We moved into a fourplex in Forest Hills, Queens, which was a more fitting neighborhood for a Cass to live. Bill would tell people we lived across the street from the Forest Hills Tennis Club, not realizing that living next to an exclusive club isn't the same as belonging. Delusions aside, Forest Hills was a good place for us. There were three movie theaters and a number of delis, a bakery, and even a Gap. Proudly wielding his new American Express card, Bill bought us a big TV, a VCR, and a piece of original art. He would stop by Balducci's on the way home from work and pick up three beef tenderloin steaks, which my mom would fry up in an iron skillet. Bill would then drown his in ketchup.

I was in junior high, and Bill was following my education closely. Appalled at how shoddy the New York public school system was, he started giving me books by Henry James, John Stuart Mill, Plato. When he learned that I had taken a liking to *He-Man and the Masters of the Universe*, he stepped up his efforts to expose me to the history of cinema. Bill had always taken me to movies that were way over my head. (Sometimes this was good, like in Seattle, when he pulled me out of first grade to go see Stanley Kubrick's *2001: A Space Odyssey* at a revival house. *A Clockwork Orange*, however, in fourth grade was less appreciated.) Now we poured over the shot selection in *Citizen Kane*, compared the different techniques used by French New Wave directors, and worked on how to cultivate the proper ironic detachment while watching *Entertainment Tonight*.

Bill also started his informal lecture series, though this was largely a function of the fact that he didn't have any friends. While I sat on the couch, squirming, Bill would pace back and forth in our living room, holding forth for hours about Roman aqueducts or battle tactics during the Napoleonic Wars. His focus was remarkable, almost chapter-by-chapter recountings of whatever books he was reading at the time, all delivered with the lexical precision of a Shakespearean actor. "You do know who William the Conqueror is, right?" he would say, and I, ever gullible, would answer no, and that would at first precipitate in him a disbelief bordering on disgust, followed by a lecture that might last an hour or two or three. If I looked bored or failed to "grasp" a concept, he got angry at me for not respecting that he had the floor.

I wish I had that kind of presence of mind, but I am more floaty. As fall turned to winter, as our second trimester became our third, there were moments when I was so frazzled I was having quasi-blackouts. I would be in my home office answering e-mail, and the next thing I knew I was in the living room, reading the label on a bottle of contact solution. With the baby on the way, I had this image of chaos: strained sweet potatoes splattered on the ceiling, diapers hanging on clotheslines strung through the house, an old-fashioned washtub sitting in the dining room. Meanwhile, I was gently laying a bag of garbage in the crib while my baby cried out in the alley.

It was around this time that I started noticing television commercials for Strattera, a prescription drug from Eli Lilly that promised relief for adult ADD, which the narrator said was "like the channel keeps changing in your mind and you can't control the remote."

One Strattera ad starred a White Middle-Class Everywoman, and you're not going to believe her day. Supermarket, traffic sign, flower, blurry crowd, mean executive presiding over meeting, bird, lightbulb burning out, an empty stroller rolling down the sidewalk, balloons, birthday cake, airport gate signs, open refrigerator, makeup counter, fiddling with a pen, one-hour photo counter, receipts, bills, blurry crowds, a friend who is out of focus, dumping out a purse, cell phone, husband asking "Are you okay?"—all presented in a buzzing fever dream that wasn't that stylistically different from an antidrug PSA. If you didn't know she was supposed to be sick, you'd think she was high.

That empty stroller rolling down the sidewalk—an image they used twice—spoke to my chief concern about becoming a father. Watching this ad, I laughed the way you laugh when someone tells a joke and you feel like they're secretly talking about you. Maybe there was more than marketing at work here. Maybe my failures of cognitive control were more systemic. Maybe I had a touch of the ADD.

I did some digging. There is no blood test for ADD. (The formal name for ADD is attention-deficit/hyperactivity disorder, or ADHD. Hyperactivity doesn't present in all cases, and the *H* is often dropped in conversation.) Imaging studies have shown decreased activity to the prefrontal cortex, a region in the brain's frontal lobes that has been associated with, among other things, "executive functions," while on the cellular level deficiencies in the neurotransmitter dopamine have been associated with the disorder, though neither of these findings have been developed to the point where they can be used to make a medical diagnosis. Instead we have a combination of clinical observation, certain specialized computer tests, and self-report. However, I learned that if I did have adult ADD it was likely that I wouldn't know it. An estimated eight million adults have the disorder, or around 4 percent of the population, but of that number only 15 percent have been diagnosed. I reasoned that I had a whopping 85 percent chance of not knowing, which to me seemed like pretty powerful odds.

I visited Strattera's website. All was calm. Mother and Child lingered on a bed in soft focus. Mother read Child a book. Best of all, Mother was not absentmindedly holding a paring knife.

There was also a six-question quiz that seemed to be cribbed from the World Health Organization's more comprehensive eighteen-question ADD screener. "How often do you have difficulty getting things in order when you have to do a task that requires organization?" "When you have a task that requires a lot of thought, how often do you avoid or delay getting started?" "How often do you feel overly active and compelled to do things, like you were driven by a motor?" After each question was a five-point scale. Never. Rarely. Sometimes. Often. Very often. I felt like I was at the supermarket, weighing a piece of fruit in my hand. I clicked on submit and the answer came back with the recommendation that I see my doctor.

What the Strattera commercial neglected to mention was that in order to have the adult version of ADD you need to have had the disorder since you were a child. You can't catch it like the flu. And you have to be symptomatic all the time, and these symptoms have to lead to an "impairment of life activities." But "all the time" is subjective, and an "impairment of life activities" can mean anything you want but can't get.

When I presented the results of my Web screener to the counselor at my doctor's medical group, she acted as if I had skinned my knee.

"Ooh," she said. "That's not good."

We were in her office in downtown Minneapolis. The room was decorated with nonconfrontational art.

"I took this questionnaire on the Web," I said. "I thought you might have something a little more professional."

She pulled out what she had on file, which turned out to be the complete World Health Organization screener. There were some familiar questions, but others were new. I tried my best not to exaggerate, and even downplayed some of my answers to compensate for latent tendencies to overreact.

"Well," she said cheerfully once we were finished. "It looks like you have ADD!"

"Really?" I said.

I wasn't ready for this. I had expected either to be told not to worry about it or hear something a little more in-depth. I hope the millions with ADD received a more thorough exam than this.

"I'm sure there are some good self-help books out on the market to help you deal with that," she said.

"That's it?" I said. I felt like I wasn't being taken seriously. *I am not a silly person!*

The counselor said that my options were to go on medication or learn to live with it. She then reminded me to stop by the appointment desk to set something up for two weeks from now. I said that I would, even though in my mind she was already so very, very fired.

I was skeptical about her diagnosis, but it was hard to dismiss an idea twice. When I presented Liz with the idea, she was supportive but wary. "You have selective ADD," she said. She smelled smoke from the excuse factory and sensed future wrigglings out of various household and baby-preparation duties. She was not entirely wrong. The Japanese didn't even have a word for depression until a pharmaceutical company introduced both the concept and the relief. ADD would give my shortcomings the legitimacy of a disorder without the stigma of disease, and I felt like I might enjoy the diagnosis. Whether I had adult ADD or not didn't matter, because now I wanted it.

I started to retrofit my past to fit the diagnosis. I remembered that my first-grade teacher noted on my report card that at times I refused to let other children talk, which seemed now to indicate impulsiveness. I sent off for my school records and then started looking for an experiment that would prove what I already believed was true.

Dr. W., a researcher and clinician affiliated with the University of Minnesota, agreed to help, provided that I not disclose the details of the Conners' Continuous Performance Test (CPT), the principal screening test clinicians use to diagnose ADD. His fear was that people would do exactly what I was about to do, which was convince themselves they had a problem they probably didn't have.

As psychological screening tests go, the CPT has a long history. The first Continuous Performance Test was developed in the early sixties by a researcher who was testing radar operators' ability to sustain attention. The types of stimuli in these tests vary—one involves staring at a clock—but every CPT is the same in one regard: it never lets up. The test's boot is always on your neck. The stimulus baits and taunts and presses the same question again and again and again: Can you deal?

When the day came for my test, Dr. W. led me to a testing room that was generic and impersonal; I could have been in any empty cubical anywhere in the world. That left my brain, the computer screen, the almost comically boring stimuli, and the keyboard. I started the test, which involved hitting the space bar on the keyboard in response to the stimuli. I went in thinking I would be defusing a mock bomb or solving some kind of rebus under duress, but instead Conners's computerized CPT isn't even in 3D. It's in 2D with cheap graphics and a standard keyboard, and it runs on a computer that seems old enough to take punch cards. I felt like I could do better myself.

Dismissing the test's appearance was a mistake. The CPT would prove to do its job very well. My lack of connection to the physical brain had tricked me into believing that my brain was either on (doing a math problem, say, or wondering whether to get the chicken or the fish) or off (a natural neutral resting state that was the neurobiological equivalent of smooth jazz). There is no natural neutral. The brain works very hard to be "off." This was what the CPT was testing.

ADD *seems* like a matter of excess excitation, but it's more a matter of a lack of inhibition. Excitation is easy. It is the brain doing what it automatically wants to do, resulting in what is sometimes referred to as the "prepotent" response. Inhibition takes a little more effort. In the case of voluntary attention, the brain does double duty, processing one stream of information while suppressing another.

In the color Stroop Test, for example, words for colors are written in a different-colored ink (the word "red" appears in navy blue) and subjects are asked to attend to and say the color as opposed to the word. But when you're presented with a green-colored white or yellow-colored orange, even if you've spent your life reading little more than the backs of cereal boxes, your brain's automatic response is to read the words "white" or "orange" rather than say the colors green or yellow.

Subjects who take the Stroop Test often show visible signs of struggling with the task, stammering and sputtering like they're trying to buy a ticket at a foreign train station, but inhibition can happen very quietly as well. In the Guilty Knowledge Test, subjects are shown a playing card such as the ace of spades. Later, while stuffed into an fMRI, subjects are asked by researchers to lie about seeing their card. Even though all the cards are hidden from the researchers, the scanner sees the truth. When people say they haven't seen the ace, there is increased blood flow to the prefrontal cortex and other regions associated with inhibition. Outwardly, someone may seem to lie effortlessly, but their brain is hard at work.

The seesaw between excitation and inhibition happens at all levels of the brain. There are contradictory forces at work at the neuronal, regional, and global level. The neurotransmitters glutamate and aspartate excite the brain, egging on neurons. Gamma-aminobutyric acid (GABA) and glycine are neurotransmitters that say it's time to settle down. The body's regulatory functions are handled by the sympathetic and parasympathetic nervous systems, which, respectively, excite and inhibit heart rate, blood flow, and respiration. On the most global level, the frontal lobes act as inhibitors on the brain's emotion and drive systems. No wonder I was always tired.

Back at the CPT, my ability to inhibit started to fail me. I began experiencing failures of cognitive control. I felt like I'd been taking the test for a week, and even though I had my face close to the screen, the environment started to creep in. This generic anywhere had detail after all, and it was a lot more interesting than I had initially thought. The fingerprints on the monitor, for example, made greasy rainbows on the glass that were fun to stare at. On the wall next to me was a poster of an open-air market in Venice that reminded me of an old job where the break room had the worst office art I had ever seen—a photograph of an upturned French horn filled with different-colored scoops of ice cream. I started smiling at the thought of the French-horn sundae, and that made me neglect the space bar, and in the CPT, as in life, once you fall a little behind you can feel completely lost.

Then I noticed the briefcases. There were three of them neatly lined up on a shelf above me. The briefcases were old and boxy and a little shabby. This, I thought, was strange. One briefcase, fine. But three? Neatly lined up in a row? If only the shelf weren't so high, I could pop up, grab one, and peek inside, but it was too far and I would miss at least three or four space-bar opportunities. But I couldn't stop thinking and wondering about the briefcases. That there were three of them, and that they were lined up in a row and on such a high shelf in what was clearly not anyone's office, much less the office of someone who would be tidy and require three briefcases. If only they weren't in my peripheral vision I could have ignored them, but they were right there, *almost as if they were part of the test.*

Everything made sense now. The demands the world put on my attention network were too large, too varied, and too intense to be captured by a computer test. So psychologists located the test in an environment that added to the testing experience. If the test alone were enough, I reasoned, then I would be in a completely empty room.

Dr. W. came in and left a validated parking stub on the desk next to me. Then he disappeared. The stub, which was, again, placed at the edge of my peripheral vision, was of no interest to me, but the yellow Post-it note attached to it was. I don't know how I could have inhibited one of those helpful little squares. I stole a glance. There were two sentences on it. Maybe three. I couldn't read them, but I think it was pretty obvious what was going on. Dr. W. already knew the outcome, as he had been monitoring my progress from a secret remote location. The Post-it said something like, "See me immediately," or "Don't worry. There's something we can prescribe."

A few weeks later I received my results. The first shock was how short the test is. At fourteen minutes it was significantly shorter than the hour I had perceived. I did well on "variability," which measures how consistent I was with my answering speed, but I was "markedly atypical" on "perseverations," which can indicate impulsiveness. There was, however, no diagnosis. In fact, it was a push. I had "an equal chance of this profile belonging to an ADHD clinical respondent or a nonclinical individual. Therefore, no suggested classification is given for this area."

This felt like a betrayal. Science was supposed to have answers. I could get maybes on my own. I went to meet with Dr. W. one last time.

When we met in the lobby, Dr. W. said it was not uncommon for the CPT not to offer a diagnosis. The subtext was that I didn't have the disorder but he didn't say this directly. We then went into the same room where I took the test and he pulled down one of the briefcases.

"The briefcases!" I said as if I knew.

"The briefcases," I said again, and when he didn't take the bait, I spilled the whole story about how I thought they were part of the test, and the Post-it, too, and as I was saying this I realized that it sounded completely crazy, and not fun crazy but crazy crazy.

Dr. W. chuckled but said nothing else. The briefcases turned out to contain paper attention tests that could be taken out into the field. The reason there were three briefcases was because there were three briefcases. Any deeper meaning was entirely in my head.

Dr. W. gave me two additional attention tests. The first, which was taken from the Wechsler Adult Intelligence Scales, involved doing mental arithmetic. The other was called Tower of London and required that I move colored blocks from their starting position to match the end position provided in a drawing. During the lulls I tried to draw him out about my diagnosis. I realized that he wasn't my doctor and that ethically, and perhaps legally, he was probably not allowed to diagnose me, so I didn't come out and ask but hinted strongly at my desire to know if I indeed had ADD. Dr. W. avoided the issue each time, but after three or four attempts he got nervous.

"I'm not going to give you this one," he said as he was about to produce another attention test. It was a deck of cards. On top was a colored shape. Because the test was being taken away, it seemed to hold the answers. Then Dr. W. explained his concerns. His affect was still professional and nice, but he was genuinely worried that someone might read about the test and then try to fake the results.

"There is a reason why all of these tests are protected," he said.

We attended to each other, looking for meaning. I could imagine tricking myself into thinking I had ADD but I couldn't imagine faking it. Then I guessed that Dr. W. was using his old brain to play out a possible endgame of this, a book published and available in bookstores, in which he was shown to have given out the answers to a test that some crazy person would then use to fake ADD—worst-case scenario, a kid who was then administered large doses of unnecessary drugs.

"I will be respectful," I said.

"I would hate to see someone use this," he said. His face grew a little tauter, his eyes a little narrower.

"I'm just indulging my curiosity," I said. "I wouldn't want any real consequences to come out of this."

I promised to disguise it all. He continued putting the briefcase away.

"So I guess I don't have ADD," I said, but again Dr. W. wouldn't give me an answer. He never would.

I left his office with the realization that it was ultimately going to be my decision whether or not I had a genuine attention problem or whether I was fooling myself. This reminded me of conversations I had had with a number of attention researchers but hadn't fully understood. Even with full cognitive control, the organ would find it physiologically challenging to be both focused and diffuse at the same time. The brain has to make choices.

"We take shortcuts without even knowing that we do," Barry Giesbrecht said. "There really is a yin and yang. You're damned if do and damned if don't. It's a double-edged sword. You can't see the tiger in the forest." He paused, and then added, "Even though we are otherwise computationally efficient."

What Giesbrecht was describing was perfectly normal. You want your attention system to be highly flexible, because it's one of the first lines of defense against your environment. Think about a prehistoric brain out on the savannah, at a time when the relative safety of city life was inconceivable. If you were building a fire, or puzzling over a rock you had never seen before, and something that wanted to eat you came along, then you would want your attention to be easily interrupted.

But this doesn't mean that attention is all reflex. Alan Kingstone, professor of psychology and neuroscience at the University of British Columbia's Brain and Attention Research Lab, designs attention experiments that try to mimic the more complicated demands of the modern world. Typical attention experiments are often absurdly simple. A subject stares at a fixation spot on a computer screen, then a box suddenly appears to the left and an fMRI or electroencephalogram (EEG) tracks the brain activity as the eyes reflexively react to the stimulus. Or a subject is presented with an arrow to cue his or her voluntary attention in a particular direction. Why bother?

Kingstone, on the other hand, shows subjects pictures of real-life situations—an art class, a basketball player driving to the hoop, a politician speaking at a parliamentary proceeding. Subjects are hooked up to an eye-tracking device and then asked to infer where the people in the picture are attending. Kingstone has found that subjects use eye cues as well as contextual information. In other words, attention is also modulated by meaning.

Voluntary versus reflexive and diffuse versus focused seemed less important in this context, which I extrapolated to mean that meaning was more relevant to attention than simple inhibition. As Dr. W. once said to me, "Kids with ADHD don't have any trouble paying attention to video games." Maybe this was why I was having a hard time planning to be a father. The baby didn't mean anything to me yet.

One night Liz and I went to Babies "R" Us. She had a checklist provided by the store titled "Must Haves!" and judging from the sheer number of items on it, you'd think we were going to colonize the moon. Shopping seemed like a true test of attention: you have focused (I want *that* diaper pail) versus diffuse (So, where are the nursing supplies?) and reflexive (New! Improved!) versus voluntary (For the next ten minutes I shall scrutinize baby wipe warmers).

Without question we were faced with challenges concerning the allocation of resources. We walked the tightrope of attention. Liz was sharp, a model of voluntary, focused attention, running down the list, weighing options, cross-referencing bits of information, reviews, advice from friends, her eyes always looking, appraising, judging, knowing. But I had a failure of cognitive control before we were out of the first aisle. My brain couldn't process the fact that there were eighteen car seats to choose from.

"Maybe you're just not interested," Liz said in a moment of semi-exasperation. She had been trying to engage me in a frank and open conversation about crib mattresses. I told her that of course I was interested. Only a monster wouldn't care about his first child's layette.

THE NUT OF EMOTION

Whether or not I had adult ADD, I was now, at the very least, aware of an entire class of unwelcome, unnecessary thoughts that bubbled up in my mind, from internal arguments with Bill, to imaginary defenses of unpublished works in front of talk-show hosts, to the careful exacting of literary revenge against staff writers for *The New Yorker*. I had always believed the world was to blame for being so provocative—the artwork for a *U.S. News & World Report* story on adult ADD showed a baby, a set of car keys, and bills swirling around a man's head as if his problems were all external—but what I had blamed on the world probably had more to do with my own problems with inhibition. I was the one who couldn't make up his mind.

I used to think the source of these thoughts was purely intellectual, but just as there is a perceived split between mind and body, there is also one between reason and emotion. Brain science was teaching me to look at emotion differently.

The science of emotion does not concern itself with subjective feelings, but with objective, measurable physiological reactions. *Emotions* are the physical reactions—increased heart rate or sweaty palms—that everyone shares in common. *Feelings* are the subjective, individual expressions and interpretations of emotions. Science couldn't tell me what someone was feeling—that, after all, was the story of their own mind—but it could detail what was happening to their body, which brought me to a cluster of neurons deep in the brain called the amygdala.

The amygdala, which is Latin for "almond" because it was originally thought to be almond shaped, is one of the most unpronounceable structures in the human brain. For months I tripped over it in conversation, saying, "Ah-*mig*-duh—, ah-*mig*-duh—" before finally spitting out "Ah-*mig*-duh-lah." Otherwise the amygdala has a good gig, as it's involved in almost any stimulus that could conceivably have emotional value. Researchers have elicited a response in the amygdala with stimuli as varied as pornography, celebrities, and whole-grain cereal, but the amygdala is most famous for its role in the fear response, which is sometimes referred to as the stress response or the fight-or-flight response.

What makes the amygdala remarkable in terms of fear is that there are few other structures in the brain that have a more advantageous physiology. Typically, sensory input winds its way through the brain, which refines the information and piles on meaning and makes associations and connections. The amygdala, on the other hand, has, in the words of NYU fear researcher Joseph LeDoux, a "low road," a direct connection to the visual cortex that allows it to act as an immediate threat detector.

The power of the amygdala is its ability to direct resources. The amygdala doesn't have any role in how you subjectively experience fear; it does not tremble and shake when someone bares his teeth at you. Rather, the amygdala acts as the boss of fear. When you run across a fear-causing stimulus, the amygdala recruits the hypothalamus, pituitary, and adrenal glands—

also known as the "HPA axis." In the face of a perceived threat, the HPA axis—by triggering the release of epinephrine (a.k.a. adrenaline) and norepinephrine—makes sure your large muscle groups get more energy, and that your white blood cells get stickier so your blood will clot more easily if you get cut. The HPA axis also dampens the prefrontal cortex and the hippocampus, which, in turn, hampers your cognitive abilities. You cannot muse idly while something is trying to kill you.

Paul Whalen at the University of Wisconsin–Madison once ran an experiment to explore how sensitive the human amygdala can be. He took multiple blocks of eight different neutral faces, black-and-white photographs of people whose expressions were pointedly blank. Whalen then salted those photographs with a handful of pictures of fearful faces and presented the entire collection in rapid succession to research subjects in an fMRI. Because of the speed of the slide show and the scarcity of emotional faces, subjects reported seeing only neutral faces. But their amygdalas were activated with the brief appearance of each fearful face, even though those faces were not consciously perceived. Their brains perceived the emotion even when they *saw* nothing.

When it comes to fear, the amygdala is also a bit of a gourmand. Researchers have successfully induced an amygdala reaction with a variety of stimuli: aversive films, odors, and tastes; threatening words; fearful faces; combat sounds; noxious heat; and foot shocks. In one study, researchers were able to get a fight-or-flight response out of subjects by presenting them with unsolvable anagrams.

Bill didn't believe in the emotional brain. He had always presented himself as a hyperrational being, and I had believed him; but by our second year in New York he was turning into an amygdaloidal mess. His sense of humor, the marveling at the accents and Ring Dings, was being slowly replaced by anger and resentment. Now nearly everyone was "incompetent," a word he spit through clenched teeth, or muttered under his breath when we were out in public. When he was in a more playful mood, and playful meant not so angry that he had trouble speaking, he would affect a resentful simper when talking about his fellow New Yorkers, as if most of them were confused children. "Inferior" was the word he used for individuals, but it was the collective that rankled him the most. He called them "the ignorant masses."

The ignorant masses were everywhere. The greatest city in the world was not, as he had thought, based on native intelligence and an appreciation for Mahler. Bill was convinced that even the old-money families and the intelligentsia acted on the whims of the ignorant masses, who blindly controlled the entire city, either through their grunting protests of discontent or their greedy, brutish desires. The worst part was that the ignorant masses didn't care that they were ignorant. They should have felt shame at their stupidity—but sometimes they even felt pride!

"It's a goddamn *newspaper*," Bill would say on seeing a *New York Post* sports-page headline: JINTS WIN. "And they are misspelling things *on purpose*!"

This amygdaloidal sensitivity might have been very helpful for Bill tens of thousands of years ago, which, as one can imagine, was not a time for poets. I could see how the amygdala would be very useful, for example, if one were being attacked by a lion. But we are not routinely attacked by animals. Aside from the time I was pelted on the F train with an empty two-liter bottle of Sprite, nothing violent ever happened to our family. None of us were ever threatened or attacked or even mildly intimidated.

True, there were places in the city one did not go. You could say the words "South Bronx" to a white middle-class person and get an amygdaloidal reaction. But we rarely strayed from Forest Hills, where the biggest danger was the overly confident kids who worked at the Häagen Dazs store.

One day we were playing hooky, going to see a movie in Manhattan. Even though it was the middle of the day, the F train was crowded. After the Queens Boulevard stop, the ignorant masses started pressing against us. We had seats, but there were enough people on all sides to require a diminished expectation of personal space and a more humanitarian view of your fellow New Yorkers.

Bill's amygdala started taking in stimuli, appraising them, and then deciding on whether or not to respond, and to what degree. The amygdala exerts a strong influence on the attention network. The lights flashing on and off, the grinding wheels, the omnipresent empty bottle of Yoo-hoo rolling around at our feet, the faces around us—like that guy there on the three-seater in front of us. Bill's amygdala was "calculating the percentages," as Whalen once put it to me. Which of these sights and sounds were meaningful? And do I need to fight or flee?

"Relax," I said.

"Don't tell me to relax," he said. "I *am* relaxed. There is, however, a, how shall we say, *Homo sapiens* in close proximity who is not fully cognizant of the fact that he is an ignoramus."

The *Homo sapiens* Bill was referring to was standing in front of him reading the *New York Post*. He was a big man in a dusty flannel shirt, a construction worker, or perhaps a teamster. The teamster had his back to us and wasn't aware, or was aware and didn't care, that he was encroaching on Bill's space. When the subway would lurch on the tracks, the teamster would rock back, swinging his meaty ass toward Bill's face. There was no contact, but it was unexpected and uncomfortable and Bill was shooting the guy very obvious dirty looks.

Bill was on a slow boil. Then a big lurch, the ass closing in on Bill's face—did it touch his nose?—and when the ass backed off, Bill had cocked an elbow as if he was going to give this living symbol of all that was wrong with the city a cheap shot in the small of his back.

"Bill," I said.

"I don't have to take this kind of—"

"Bill," I said.

"You don't understand what I have to deal with every single—"

Right then the teamster turned around to see what was happening, the *Post* rustling in his hands. He had a big smooth face that showed little emotion other than slight annoyance at being distracted from news of his Jints.

Bill averted his eyes. When he looked back, he viewed the teamster with forced neutrality. Bill never subscribed to the un-written rule of keeping his eyes away from other people. He was constantly checking what people were doing, and if he was caught he would play it off as if he were presiding over the car with aristocratic calm. Bill gave a gentlemanly nod. The teamster didn't react to this courtly behavior other than to hold his gaze for a moment to make sure there wasn't any more coming. When he looked away, Bill breathed out. Then he turned to me, his lower teeth bared. Bill was doing what you do on the subway when someone bigger and stronger than you is hogging space: you save up your hatred and take it out on someone else.

• • •

What kept Bill in check? My research showed that the prefrontal cortex (PFC), along with other regions like the anterior cingulate, worked together to regulate the amygdala, but not always to great success. In my ideal brain, the two systems would be equally equipped; however, the deck seemed stacked against the PFC. In her paper "The Biology of Being Frazzled," Amy Arnsten, professor of neurobiology at the Yale School of Medicine, called the PFC "exquisitely sensitive to its neurochemical environment."

The PFC is a major part of the neural substrate for the human mind (including, as I had already learned, attention) and regulates thought, behavior, and emotion. In a phone interview, Arnsten told me that under nonstress the PFC regulates much of the brain, and does it well. "A lot of what it's doing is inhibitory," she said. "Under stress this shifts. Then the amygdala seems to take charge of the brain. It does so by activating norepinephrine and dopamine, which shuts off the PFC so it can no longer inhibit and lets the amygdala be in all its glory."

It seemed to me that the amygdala had another kind of privilege, shutting down what interfered with its power with a kind of knee-jerk disregard for the consequences. When the amygdala achieved its electrochemical dream state, it not only got to inhibit the PFC, but it also created an environment under which it performed better, perpetuating a vicious circle of less self-control and more emotion.

"The amygdala gets its cake and eats it, too," said Arnsten.

"The more I learn about the amygdala, the less I like it," I said.

Arnsten, however, thought the amygdala was perfect. Her feelings toward the amygdala echoed those of other researchers, and reminded me of that old computer programming joke: It's not a bug; it's a *feature.*

"May it ever be so," she said.

There is a scene you encounter once in a while—I've seen it in as disparate places as the film version of John Irving's *The World According to Garp*, Martin Amis's novel *The Information*, and ABC's *Desperate Housewives*—of the Grown-up Furious at the Speeding Car. Usually the action takes place on a bucolic suburban street in America, although Amis stages his moment in London. Before the offending car arrives, all is quiet and calm. Then, here it comes, zooming past, blowing through the stop sign or taking a recklessly wide turn at the intersection. Our hero— usually a man, but women are allowed this moment as well— charges down the lawn in a primal rage. But by then it's too late. The car is gone, and he is left only with exhaust and indignation.

Until we got pregnant I always thought all this impotent fist-shaking was cheap shorthand for a parent's caring. Then it happened to me. I was outside my house on my bucolic city street when a car—long, low-slung, rusty—flew by at a rate my expectant-father-addled brain perceived as nearing supersonic. My attention system attended: a speeding car, with spoked wheels in a menacing spin; a suspension system that made a hulking, bottoming-out sound; and on top of it all, the airplane roar of its V-8. Next, the physiological reaction, the quickening of my heartbeat, the tensing of my leg muscles, the tightening of my skin, the hungry gulps of air. Then, as the car faded in the distance, the subsiding of the physiological reaction and the arrival of higher-level interpretations as my mind tried to make sense of what happened. A flurry of voices all talking at once: *What kind of car was that I can't believe what assholes and was that a crack pipe you were smoking while you were driving with your knees?*

Once the moment passed, my breathing returned to normal, and my vision returned to the default setting you would expect when standing on your front porch on a cool fall day while checking the mail. But still, underneath, the unspoken fear: concern for my baby's basic survival. That question still hadn't gone away. Can we do this? There was no way to know. People said we'd be fine. And we probably would be. But you don't *know*.

I am not macho. I am not proud. But it was important to me with the baby approaching that I not be angry and afraid. I still wasn't sure whether or not I had adult ADD, but by now I was over wanting anything to be wrong with the brain. Quite the opposite. I didn't want to believe in the strength and sensitivity of my amygdala. I wanted to prove that my brain was tough, calm, reliable.

Melinda Miller, a second-year graduate student at Rockefeller University, was going to help. I visited her at a lab on the Upper East Side at the Cornell Medical Center. It is always strange being back in New York, even as an adult. Even when I'm in a good mood, I often feel like the city is going to leave me shirtless and dirty and begging for busfare home. This generalized anxiety seemed appropriate, given the reason I was here—an experiment in fear conditioning.

Fear conditioning has two simple components. First, there is an unconditioned stimulus, in this case an electric shock. The stimulus is "unconditioned" because no one needs to teach you that electricity coursing through the nerves of your wrist is painful, or at least annoying. The second component is the conditioned stimulus. This can be a tone or a light, anything that would otherwise not be associated with any emotional reaction, either positive or negative. Today my conditioned stimulus was going to be a blue square.

The genius of the brain is how the amygdala pairs the two stimuli. The physiology at work is simple and elegant. Imagine that one neuron handles the blue square, happily firing whenever blue squares appear, but remaining otherwise quiet. Another neuron handles the shock response, firing, with equal happiness, whenever an excess of electricity in your wrist triggers the fight-or-flight response. When both shock and blue square appear together, the amygdala plays matchmaker, tweaking the shock neuron so that it now also fires when it sees blue squares. After conditioning, supposedly all I would need to do is see the blue square to have the fear response—not the full horror-show effect, with trembling skin and voided bowels, but a measurable physiological response (increased heart rate, increased respiration, a tightening of the skin, increased perspiration). The experiment is so robust and universal, it's gotten results with humans, monkeys, dogs, cats, pigeons, worms, and even flies.

Miller prepped my fingers, first with NuPrep exfoliant, then with a soothing layer of Signa Gel conductance gel. She applied finger sensors that would pick up the downstream physiological changes as directed by my amygdala, and then she taped the shocker bar to the inside of my right wrist.

"Before we start the experiment, we'll have to set the shock level," said Miller. "It has to be something you're really not looking forward to, but we don't want it to be at a point where it's painful."

"That machine goes up to painful?" I said.

"Everyone has a different threshold," she said. "But there is an upper limit that I'm not allowed to go past. If you want a comparison, most electrical outlets are a hundred and twenty volts. The most you'll get is about half that."

"Half an electrical outlet," I said, as if this were some legitimate unit of measure.

"For rats it's a lot scarier," she said. "We can't talk to them, so they all get the same intensity of shock."

The shock box started at ten volts and went up to sixty-five volts. Miller set the shock and fired a test. It elicited a mild twitch, but was otherwise bland and uninspired. Two stars. We try again.

"Gah!"

My jaw clenched involuntarily. My hand jumped. My fingers felt like they were made of needles and hot sand, and the current went all the way up to my jaw.

"I can move it down if you want," Miller said right away.

"Oh, man," I said.

"I can *totally* move it down," she said. "If it's painful, then it's too much."

I had already forgotten how much it hurt. I asked her to try again.

"Guh!" She must have this thing pegged at sixty-five.

"Want me to move it down?" said Miller.

"I'm not trying to be a tough guy, I just want to be sure the test *works*," I said. "Where am I at, by the way, just out of curiosity?"

"You're at thirty."

"Oh," I said. "So I'm about halfway?"

There was a polite silence.

"Some people just have sensitive skin," she said.

Miller then wheeled over to her laptop, which rested on a desk across the room. With a quick keystroke the experiment began.

When the first blue square came up I knew this was potentially a shock square. Then the shock came. My teeth clamped down, my arm jumped. I felt a kinship with all the rats, pigeons, and flies that came before me. Then a second blue square and all the skepticism in the world didn't matter. I was bracing for that shock like it was a punch in the face. The internal monologue ramped up, the human equivalent of "defensive behavior" in rats. I could feel my body reacting. The next stimulus, a yellow square, was a safety color, the control. No shock here no matter what. My face grew soft. My body slumped a little. I was filled with experiment shame. I didn't need to see the data to know that I had conditioned in one trial.

For the rest of the experiment every blue square yielded a keen amygdaloid reaction, even though I wasn't shocked more than two other times. As it turned out, I didn't need much to have the fight-or-flight response. Thirty volts. That was me. That was my breaking point. One quarter of an electrical socket, for anyone who has future plans to torture me.

"It looks like you definitely conditioned," said Miller. We looked at the data on her laptop, a blue squiggly line dancing across a graph. Bisecting the squiggle line were vertical lines that indicated where the colored squares were presented; about a half-second later there was the response. For the yellow squares, there was no reaction; for every blue square, a big jump.

The amygdala reaction was almost instantaneous, but the skin conductance took half a second to register. Even with the lag, the meaning of the graph was clear. The jump was in two parts, first the startle of first seeing the blue square, then little upticks of worry as I wondered whether or not a shock would come. During the experiment, I didn't feel all these discrete moments, but now this further refinement made sense. Wasn't this the nature of anxiety, the slow accumulation of matchsticks of fear?

Strangely, the graph didn't show a bigger reaction after the shock, after the actual pain. Miller said that different people had different galvanic skin responses to the shock, but my personal data reflected other research that has shown that the amygdala is not involved in pain. Subjects exposed to "noxious heat" versus a "warm" stimulus did not show a greater amygdaloid reaction for the pain. It didn't mean that worry is worse than pain, but more that the amygdala finds worry more interesting. For the amygdala, fear has more texture, more information, more meaning.

Miller summed it up right before I left.

"You'll never look at a blue square the same way again," she said.

I left the hospital and started walking through the Upper East Side. I was seeing blue everywhere: the *New York Times* vending box, the MTA logo, the Agra Indian restaurant sign. My jaw felt like I'd just spent two days straight chewing the same piece of gum. I now knew that fear worked with a richer and broader palette. There were opportunities to condition everywhere, and once the fear and the normally neutral stimulus were paired, there was reason to believe that the marriage was going to last. "A strong case has been made that long-term memories for Pavlovian fear conditioning, at least those that support fear CRs [conditioned responses], are mediated by synaptic plasticity in the amygdala," writes Stephen Maren in his paper "Building and Burying Fear Memories in the Brain." "By this view then building fear memories requires long-term synaptic modifications within the amygdala. And given that the amygdala is involved in maintaining Pavlovian fear memories for almost the entire life span of a rat, it appears that these memories and their synaptic correlates reside in the amygdala *permanently*." (Emphasis mine.)

Miller downplayed the fact that I had a good-sized conditioned response. She said it could be because of my naturally high skin conductance or because of humidity in the room or my "sweat gland genetics." But I wasn't sweaty. Sweaty wasn't the reason. Researchers have found a correlation between the size of the amygdala response and the size of the conditioned response. I took this to mean that I have a more sensitive amygdala than my conscious self would like to admit.

I found the shock experience both unsettling and clarifying. For one, I felt like I needed to revisit my New York fear history. I used to think in terms of autobiography, how I was afraid to play in a basketball game in high school because we were getting beaten so badly. Or how I was afraid of going to school when we moved to New Jersey to the point where I didn't want to take off my parka in class, as if it were going to protect me.

I had left New York and lived in Minneapolis in order not to be afraid, but now I learned that I had taken my amygdala with me. There were fears that I didn't know I had, hidden associations that were remembered at the neuronal level, far out of reach from my consciousness—a lifetime of blue squares. According to some researchers, any firing of the amygdala that reached a certain threshold could be considered a fear response, even if the organism didn't subjectively feel the fear. If I hadn't been paying attention to the minutiae of my response, I might not have even noticed the subjective experience of fear, but then the amygdala didn't need to send you running and screaming in order to shape your life. Just as Bill didn't have to know he was afraid on the subway for it to affect his behavior, I didn't need to feel afraid for fear to invisibly rule my brain.

At the time, I never understood how Bill could be angry at New York for being New York. It seemed to me that if he didn't like it, he had the choice to either accept the situation or leave. There would have been nothing to stop him from saying this was a failed experiment and retreating, taking that big-city résumé back to Seattle and getting a job as a VP at a bank. But now I saw that Bill probably didn't know he was afraid. The brain doesn't have an internal diagnostic system to tell the self that the organ is out of whack. There is no engine light.

I didn't see the connection, but my growing anger wasn't all that different from Bill's. I was starting to feel betrayed by science. Given my experiences so far, I felt that the popular neuroscience I'd fallen in love with was getting it all wrong. Popular neuroscience treated the brain with an overabundance of wonder and awe and reverence, as if the brain were solely devised to appreciate Mozart and Shakespeare and the poetry of e.e. cummings. Why were we always kissing the brain's ass? You would never know we were capable of a misguided war, or that the sugar industry was destroying the Everglades, or that there was a sport called "competitive eating." The bad brain was always small and far away, as if viewed through the wrong end of a telescope. Then there were the enduring myths, such as the idea that we only use 10 percent of our brains, which is nonsense. We're using all of the brain all the time, both halves at full speed. Look around: this world, this life, is the product of the human brain at the top of its game. In those closing days of 2003, with our baby due any day, even the world's most perfect brain didn't seem like it would possibly be good enough.

I BLAME MY PEPTIDES

We spent the last hours before our son was born frantically trying to install a ceiling fan. It was five in the afternoon, which on a January day in Minnesota meant that when we turned the power off for the sake of safety, the entire first floor of our house was steeped in blackness. I stood on a yellow stepladder in our living room. Liz, a week past her due date, trained a flashlight on the hole in the ceiling, but then a contraction came and the light disappeared.

In complete darkness, my arms shaking from holding the fan over my head, I listened to an assortment of grunts and groans. I waited. Then the light came back and I continued my absurd, frantic dash to complete a project that probably could have waited until after the baby came.

"Sweetie," I said, "we don't have to do this now."

"We're going to finish it," Liz said when she came up for air. I was looking into the flashlight, partially blinded. It was pitch-black, but her voice was bright with possibility.

I worked on the tangle of wires and the stubborn electrical box but then the light went away again, this time accompanied by a louder groan. When the light came back it went away again almost immediately.

"Liz, honestly."

"Finish it," she said.

The next time the light came back for such a short period of time it was like someone had lit up the ceiling hole with a strobe. I went on strike.

"Liz, there is no way this can't wait until—"

"Finish it!" she said, and her voice became hoarse and un-recognizable. *"Finish it!"*

I thought about her amygdala, but I didn't say anything. When your wife is in labor it's best not to say it's all in her head.

I felt guilty for thinking of her in these terms, as a cluster of neurons and not as a person, but this was where I was with the brain. I had become interested in the brain to find out how my own wits worked, but whatever knowledge I had absorbed was turning against me. I became obsessed with the thought that it was essential that we overcome our commonsense dualism. If the brain were aware of itself physically, then it could signal that something was wrong. One scientist I talked to about this strongly disagreed. He said it was like having bad eyesight. "You just want to see," he said. "You don't want to be aware of your glasses."

He was right. But it was too late. Now I saw the brain every-where, even in situations like these, when I needed to be thinking about people as people and not as a collection of electrochemi-cal signals. Whatever Liz was experiencing went beyond the in-ternal war between amygdala and PFC. I was oversimplifying and was reminded of how popular neuroscience articles often mistakenly talk about how the brain is "wired" a certain way. Even if a neuron loosely functions like a wire, the brain is far more complicated, and far less linear, than an electrical circuit. The brain is a complicated mess of neurotransmitters, neuro-hormones, and who knows what other kinds of pan juices. Liz's amygdala was only part of her reaction, and this was especially true when it came to the neurobiology of being a parent.

C. Sue Carter, codirector of the Brain-Body Center at the University of Illinois–Chicago, is one of the pioneers in the study of oxytocin and vasopressin, two peptides (amino acid proteins) that are thought to be crucial to parenting, mating, and social interactions. Carter got interested in these two hormones when she failed to find the answer to social attachment by looking at sex hormones like estrogen and testosterone. "When I first started all this I did it using sexual preference," she said in an interview. "Even paired animals would mate with strangers, but they wouldn't sit next to them or touch them. Instead of having what I expected was a strong preference for who they mated with, they showed a strong preference for who they lived with."

Between oxytocin and vasopressin, oxytocin gets the most attention. In popular science articles, it's often called the "love hormone." You probably won't get a love high by injecting oxytocin into your bloodstream, though if you're a mammal you will start acting like a mother. Infusing a non-pregnant ewe's brain with oxytocin will stimulate her to bond with an unfamiliar lamb; injecting oxytocin directly into the brain of a virgin female rat induces full maternal behavior almost immediately. Oxytocin also inhibits the stress response. Rats that have been treated with oxytocin show less anxiety in the Open Field Test, which probes their level of comfort with novel environments. Male rats treated with oxytocin will also spend more time being social with other rats, which isn't usually their thing. Perhaps most telling, the administration of opium to an infant rat will not suppress its separation cries, while the administration of either oxytocin or vasopressin will.

Oxytocin is released in women during childbirth, and oxytocin is also passed on to the child during breast-feeding. "Women get the double whammy," Carter said. "But men get affected, too."

But you have to be there, which is maybe why Bill, as a stepparent, took a while to bond with me. Arriving on the scene when I was three, he missed out on the chance for whatever oxytocin whammy I could deliver. One of Bill's favorite stories was about the day he realized I was interesting. I was six or seven years old; we were still in Seattle and Bill and my mom were in their party years. My biological father came over for a visit, and apparently things got a little awkward. Bill said that suddenly I started talking, telling stories and making observations about the world that impressed Bill as delightful and precocious. When Bill told this story later, he acted out his own eureka moment. "I thought, *Wow*," he said, "maybe this is a person who might be worth knowing."

Later, Bill legally adopted me. On a cold and rainy day a few months before we left New Jersey for Queens, we drove quietly to Bergenfield, New Jersey, and met with a judge who asked me if this was okay with me. I told him I supposed it was. Bill had approached me with the idea only a few days before; he seemed torn between seeing it as a formality and something I should be deeply grateful for. Afterward, Bill was pretty happy with himself, and maybe he got some surge in oxytocin now that he was legally my dad. If he did, he didn't show it, other than to take me out for pizza.

When my son Owen was born, on the other hand, I got the

full whammy. His was a long labor and his umbilical cord was wrapped around his neck, and he had been squeezing it in his hand, and by the time he came out he was so purple he was almost black. My brain was in the room for an instant, as I wondered briefly how my amygdala was reacting to the experience, but then the engine light went off and I was a person in full, stroking Owen's little head with the tip of my index finger while he trembled in his plastic tray.

Later, when I talked to Sue Carter about my experiences as a new father, she called a newborn baby "an endocrine manipulation." (This was not entirely out of character. In her paper "The Neurobiology of Love" she describes physical closeness as "the maintenance of proximity or voluntary contact with an attachment object.") While this sounded a tad clinical, I also understood what she meant. Even if I hadn't been studying the brain, there was a distinctly chemical feeling to the flood of love I felt for this boy.

I could see where the idea for Jesus came from. James Leckman, in his paper "Early Parental Preoccupations and Behaviors and their Possible Relationship to the Symptoms of Obsessive-Compulsive Disorder," notes that 73 percent of mothers and 66 percent of fathers say that at three months their child is "perfect." There was that moment in the hospital when I looked at Owen and thought, *He's the son of God.* There was no way to look at him and not feel that he had been sent to this Earth by a supreme being to right the world. This was what Joseph and Mary had experienced; no one had bothered to tell them this was how most new parents felt.

This feeling of the perfection of the universe extended to Liz as well. I had never felt more love for her, or more of a sense that we were meant to be together and start a family. But the joy soon manifested itself in unsettling ways, too. During those first days in the hospital it was very important that everyone agreed that my son was cute. And not only did they have to agree that he was cute, but in my newfound effort to be scientific about my life, they had to agree that their high appraisal of his beauty was an "objective measure."

"Cutest baby in the room," said the pediatrician who was on rounds.

"I know," I said. "But, really. You're a doctor. Don't you think he's the cutest? I mean, objectively."

"Cutest baby in the room," he said.

We drove Owen home slowly, as if transporting a load of nitroglycerin through the jungle. When we arrived, Liz went inside alone, leaving me to unhook Owen's car seat from its base, but I was feeling so euphoric, not to mention unprepared, that I didn't know how. So I closed the door and stood in the cold and snow, reading the instructions while he was stuck in the car. After I had been blissfully trying to figure out the car seat for fifteen minutes, Liz came back out of the house. "You've got to be kidding me," she said from the front porch, but with none of the high anxiety of the ceiling-fan moment. She sounded almost glad that I had my head up my ass. We looked at each other and smiled. Everything was natural and true.

As we went through those first few weeks, the manipulation of my endocrine system continued and it seemed to carry over to everything I did. It was like I was suddenly some kind of hippie. Oxytocin is partially mediated through the sense of smell, and I would huff Owen like he was a bag of paint, then set him down and enjoy everyone and everything.

This raised a question about these social-bonding peptides: Do they affect other aspects of our behavior? Sue Carter is looking to explore how these endocrine manipulations can lead to more generalized behaviors. The presence of a prairie vole pup will cause a rise in oxytocin in both male and female prairie voles. The pups act as their own self-contained oxytocin injection. Carter wants to see if these boosts in oxytocin affect the social behavior of voles toward other voles and not merely toward their mates and pups. The other question, and one that is harder to answer, concerns what secondary effects oxytocin has on memory, learning, and other higher functions. The voles that are used in most attachment experiments don't have much of a cortex, which makes it hard to use the animal model to support those kinds of investigations. As Carter remarked to me, "They don't do well with mazes."

From the very beginning I had kept a diary to help me track the thoughts I was having about the brain. In this "Brain Log," I asked myself questions like "Is there a sub-subconscious?" and "What is the neurobiology of indie rock?" and kept lists of everything I ate and drank, as well as all the media I consumed in an attempt to see if there was a connection between these factors and my general mood. Mostly I filled the diary with ramblings and half-baked ideas. (I was kidding about these thoughts but also not kidding. I used to joke with my friends about how Keanu Reeves was the greatest actor of our generation. I have said this so many times the joke has disappeared. I now believe it's true.) There was Small Hole Theory (the brain is like a clarinet: raw air/data goes in one end and what comes out depends on which holes are covered), Wily Brain Theory (the secret to the brain's success is its unwillingness to be understood), and Nowture (neither Nature nor Nurture but a theory of behavior based on what is happening *right now*). I was the smartest man on earth and I didn't have to leave the house.

After Owen was born I tried to be scientific about what I was experiencing. I created Thought Logs, data-collection sheets that I tended to and worried over and adjusted and manipulated, generating version after version. Each had a system of 1-through-10 scales, with categories like Alert, Mood, Focus, Stress, Output, Creativity, plus space for little notes off to the side like, "mild frenzy at 4:30 p.m." Between 9 a.m. and 1 p.m. I thought about "conquering the world," though I failed to write down how or why. Some entries were a complete mystery. I don't think I'll ever know what I meant by the phrase "medium horny."

I was starting to fancy myself an amateur neuroscientist. Frustrated and disillusioned by what real science was telling me about myself, I thought I could strike out on my own. I boned up on the Scientific Method, read Thomas Kuhn's *The Structure of Scientific Revolutions* and *Advice for a Young Investigator* by Santiago Ramon y Cajal, a Spanish biologist who is considered the father of modern neuroscience. The whole endeavor looked pretty easy. I only needed to observe a phenomenon, create a hypothesis, create an experiment to test said hypothesis, and then use the results to revisit my original theory.

I started keeping notebooks of ideas for experiments.

> Idea: Memory walk where people go for a walk and then we record what they remember and don't remember from the walk they just took.

Idea: Give people money to spend and use that as an example
of inhibition and excitation.
Idea: Have people observe my behavior when I'm hungry.

I had also generated a list of seminar titles that contained
what I believed were kernels of legitimate investigations. I had
these fantasies that I was going to present my findings at the Society for Neuroscience's annual meeting in San Diego later that
year.

What Are You Thinking?: Saving Modern Neuroscience
from Itself
That Was Easy: New Theory Solves the Brain
Why Didn't I Think of That?: Why You Didn't Think of
That

In the shower, when I would practice my acceptance speech
for the Nobel Prize for Neuroscience, I imagined the surprise on
the community's face when the word went out. An English major
cracks open the riddle of attention while simultaneously helping
further the methodology for teaching neuroscience to under-
graduates. I would be gracious: with the accolades come respon-
sibility, and I would show a fatherliness that would demonstrate
that I was anything but petty and small.

To further this new career, I recruited the Neuroscience Club
at the University of Minnesota as research subjects and took
them to the Mall of America in Bloomington, Minnesota. My
experiment involved something I was going to call "mass attention" or "group attention" or maybe "social attention" or maybe
something else. The gist was, we were going to walk through the
mall and write down what we saw. Then I was going to buy them

lunch at the food court, where we would have a lively and informative discussion about the neuroscientific significance of what we had just done. Six subjects, an amateur neuroscientist, and one of the world's largest indoor shopping experiences simply had to add up to something.

According to plan, at 10 a.m. on a Saturday morning we met on the other side of the elevators from Legoland. Because it was extremely important to me that this experiment look and feel like a real experiment, I had my subjects sign release forms and promised them confidentiality in the event that I published my data. Next, I explained the experimental protocol they would be facing.

"There are no wrong answers," I said. "Just keep the pens moving. If you want to see shapes, see shapes. If you want to see people, see people. Not that I want to prejudice you. Although maybe I already have. Hmm. In any case, I don't want anything. There is no right answer. Just observe and record."

Research Subject MOA1 smirked at me.

"Just have fun," I said. "You may not learn anything today. But I will."

We then went upstairs and fanned out along the balcony that overlooked Camp Snoopy, a miniature amusement park set in the creamy center of the Mall of America. My six research subjects were each armed with an Ampad Reporter's Notebook (seventy pages, Gregg ruled) that I provided at the beginning of the trial. Their backs to the South Food Court (Johnny Rockets, Panda Express, Minnesota Picnic), each of them started busily scribbling away, taking notes on the scene in front of them. We were lit by a mixture of weak winter sun and artificial light. Roller coaster, Ferris wheel, mom negotiating a tantrum being thrown by her secondborn, log flume, giant Pepsi ball, teenage girls pretending that they're trouble, Knott's Berry Farm game room, abandoned mine shaft ride, obese man eating caramel corn—it was a feast of voluntary and reflexive attention, plus a full buffet of emotional stimuli for the amygdala.

I felt fantastic. Running my own lab was much better than piggybacking on other people's science experiments, tapping fingers, or listening to a list of aversive words, or staring at a fixation cross and waiting for current. Look at us! Out in the world! Doing science!

I was writing, too, serving as the control, and after about five minutes, maybe seven, I told them to stop. They flipped their notebooks shut, turned away from the mall, and faced me. I loved these kids! They were young—most of them hadn't even declared their major yet—and I felt a welling of pride. They were the future of science, and I couldn't help but feel that I was helping shape that future.

My subjects, on the other hand, were a little flat. At the time I thought this was strange, because when I'd recruited them a few weeks ago at their club meeting, everyone was enthusiastic, laughing at my jokes and offering ideas for experiments. Today only six showed up, and they were even more listless than you would imagine for a bunch of college students. Research Subject MOA2 had barely said a word.

They were simply not as oxytocin crazed as I probably was. And this made me want to bond with them even more. My oxytocin level may have added to my overall bonhomie, but to my research subjects I was just a weird guy leading them around the mall. Their oxytocin levels, I imagine, were appropriately low. Was this hindering attachment?

"I feel like I need to entertain you guys," I said.

As we headed to the next stop, which I picked at random,

Research Subject MOA1 was on her cell phone, laughing at me. MOA1 had been laughing at me a lot. Meanwhile, Research Subjects MOA3 and MOA4 weren't showing the kind of leadership I expected. MOA4 was strangely quiet, especially since in previous conversations MOA4 had shown enthusiasm for the venue I had chosen for the experiment (it was good, he assured me, for checking out girls). I wished they were having as much fun as I was, but then again I was having more fun than I'd had in a long time.

Next we stopped at a series of kiosks called Bolivian Imports. There were piles of multicolored sweaters and hats and scarves. Normally I would have passed by a stand like this, but the colors made me happy. I used to hate coming here, the mall giving me a special kind of lethargy that only 2.5 million square feet of retail space can deliver, but this time I wanted to hug the whole place. Plus, the place was crawling with babies. I couldn't stop looking at them, in their strollers and detachable car-seat carriers and Baby Bjorns. My subjects started to write. I got to work on the control, but I was having trouble concentrating. I felt this overwhelming desire to make sure my subjects were happy, especially Research Subject MOA2, who looked like he could use a boost.

Then I heard a noise. It sounded like a baby crying, but it turned out to be one of the women working the booth.

"Are you the teacher?" she asked.

"Sure," I said.

MOA5 cracked up at this. The woman explained that she was from Peru, selling Bolivian sweaters with her friend Lucy from Ecuador. I didn't know what to say to this. I was friendly but I also wanted her to go away. In part I didn't want her to spoil my subjects' objectivity. Control is the operative word in any good experiment, and I couldn't have an outsider influencing the outcome of the trial. But I was also feeling protective. These were my kids, not Lucy from Ecuador's.

There was another peptide at work: vasopressin. If oxytocin is associated with women, vasopressin is the "male" peptide, even if men and women have both. Vasopressin acts like the sympathetic nervous system, speeding things up though it can also have a calming effect, depending on the circumstances. According to "Oxytocin and Vasopressin as Candidate Genes for Psychiatric Disorders" by Emory University's Larry Young, vasopressin "stimulates social communication in birds, frogs and hamsters, stimulates aggression in hamsters and increases affiliative and paternal behaviors in voles." It has been shown to facilitate "mate guarding." Male voles given vasopressin treatments during early development were more aggressive toward strange voles.

Once I had defended my pups, I took them to the Rotunda, where there was a patio show two stories below. There were grills and barbecues and Astroturf. There was a big goth girl with cornrows and all kinds of straps on her black outfit. Then on to the North Food Court. People were staring back and giving us dirty looks as if we were invading their space by writing down how they looked when they ate. One man, a maintenance worker, turned his chair around and gave us his back. But these kids, these terrific kids whom I only wanted to take care of and make happy, kept on writing. I couldn't wait to see their data.

I had promised lunch and I intended to deliver. At the food court I passed out a ten-dollar bill to each subject. Research Subject MOA1 smirked at me again, then headed off with a few others to the Panda Garden. I went to the A&W Root Beer stand, navigating my way through more parents and babies, startling now and then when one cried. When I came back, they had gathered around two tables pushed together.

"Nothing says science like a cheese dog, right?" I said. Silence. I didn't want to be critical, but this wasn't a very fun club.

Then I opened up the discussion. MOA1 reported seeing people buying things. MOA2, who had been very, very quiet, reported seeing more bright colors than people. MOA5 was the most philosophical of the bunch, wondering expansively about the interplay between color and shadow and meaning. Then we covered each individual location. At Location #1 everyone saw the roller coaster. "I wasn't intentionally trying to see it," said MOA1, which pointed to her inability to inhibit certain stimuli. MOA3 and MOA5 both saw the same kid having a temper tantrum. Otherwise there was little overlap (although everyone agreed to a robust response when they saw the Gothic Princess at the patio show).

My theory was that there would be certain stimuli, especially emotional stimuli, that would be enhanced in the social setting. Everyone would see the temper tantrum because it had emotional value. I'd have to check the data before I formed any more conclusions.

Research Subjects MOA2 and MOA4 noted they saw a lot of baldness. (Both subjects were male.)

"I'm calling the journal *Science*," I said. "I don't know what it means, but I'm calling them."

Crickets.

I still felt great love for them, but this lack of responsiveness was starting to get to me. A free lunch ($10 value), a field trip, and a chance to do cutting-edge science—I didn't know what else they could possibly want. But I couldn't ask the reason for their distant behavior; it would have upset the delicate balance between primary investigator and research subject. My lab didn't have a formal internal review board, but if it did it would frown on meddling. Still, I had to know why they were so reticent. Finally the answer came. When I divulged my interest in visual attention, MOA1 confessed to thinking the whole thing was secretly a behavioral experiment. The subject thought that I was using the day as an excuse to spy on them.

"You thought I was trying to trick you?" I said.

Around the table, polite nods.

"In a nutshell," said MOA5.

When I came home I felt like I had failed, but I wasn't ready to understand why. Later, when I would try to make sense of that day in terms of all the many facets of the brain, it still didn't make sense. What had I been thinking? What was going on with my hormones? Could you even capture life through science?

"All this stuff you're trying to put into words wasn't designed to have words attached to it," Sue Carter once said to me. "It was designed from the bottom up. We're limited because we can't tell our problems apart. We don't know if we're in love or we're hungry."

This is why we have data. Unfortunately, when I returned to my Mall of America experiment a few months later, the data made absolutely no sense to me. My research subjects' notes were so scarce and so vague that I could barely tell where we'd been. (The exception was MOA1, who wrote these miniscenes that included a running commentary about the ills of American consumerism.) What they did write down didn't match up with our lunchtime chat. No one had talked about the big Pepsi ball, and yet everyone except MOA1 wrote it down, and yet, while the Gothic Princess was the subject of much conversation, no one had taken note of her in their notebooks. This couldn't be right.

Fine: I can always come up with another angle. I put all the notebooks in a box, put them in my office closet, and cracked open my Ramon y Cajal. I had been reading a chapter called "Diseases of the Will," his take on science's "illustrious failures." I read about the dilettantes, megalomaniacs, and "theory builders" who will never succeed in the field. Ramon y Cajal couldn't have spoken more clearly to my faults and flaws ("Basically, the theorist is a lazy person masquerading as a diligent one"), but I still didn't see it. Instead my lab stayed open for business for a while longer. I wasn't finished with bad science yet.

THE BODY SAYS NO

If a newborn could inspire the idea of Jesus, then a four- or five-month-old was where people got the idea for vampires. Owen still smelled good, and his little legs, soft as hot dog buns, always brought a smile, but the newborn hit didn't last. Liz and I experienced a kind of permanent exhaustion that often left us shaking and weepy by the end of the day. I was stressed beyond anything in the old days of work pressure and deadlines, when the word "busy" was a piece of social currency. I was not a disembodied, transcendent soul occupying an earthly vehicle but a body, an aching, trembling body.

I was experiencing a disruption of what biologists call "homeostasis." The idea is that for many of the body's systems there is a range of acceptable levels, and the brain is always comparing that range with external or internal conditions. This might be a matter of the amount of water or nutrients in the body or the level of glucose in the bloodstream. Walter Cannon, one of the first advancers of the concept of homeostasis, used the phrase "the wisdom of the body" to talk about survival benefits of homeostasis. When conditions are outside acceptable tolerances, the brain triggers a response that can be genetic, chemical, or behavioral to get the balance right. Stay out of balance for too long and you have what is colloquially known as "stress."

Whenever the body encounters a stressor, the HPA axis first produces epinephrine. This is the "adrenaline rush," the short-term juice for all those downstream responses I experienced during fear conditioning. But if the stressor lingers—if the shock continues, or if the animal keeps attacking and attacking and attacking—then the body needs more energy to keep up. This is when the adrenal glands (the third link in the HPA axis) kick in, producing cortisol. While epinephrine acts quickly, cortisol acts over time. It's the pasta instead of the candy bar.

Like oxytocin and vasopressin, cortisol also acts as a neurotransmitter. In small amounts, cortisol is beneficial. Sue Carter's paper on love notes that "male Prairie voles [form] new pair bonds quickly after corticosterone [an adrenal hormone similar to cortisol] injections." One of Leckman's parenting papers says that "first-time mothers with high levels of circulating cortisol are better able to identify their own infants' odors" and that the "level of affectionate contact with the infant (affectionate burping, stroking, poking, and hugging) by the mother was associated with higher levels of salivary cortisol."

Too much cortisol, however, as well as high levels of cortisol over extended periods of time can be more devastating than all the affectionate burping one could ever counter. Neglected rats—those that receive less licking and grooming, usually as a result of some kind of bond-breaking separation from their mothers—have an excessive stress response that lasts through adulthood. Infant monkeys whose mothers were deprived of food have offspring that are more troubled and stressed. Over time, elevated levels of cortisol also causes mineral loss in bones, and abdominal obesity, and disrupts the immune system.

In a cruel evolutionary joke, the structure responsible for shutting down the stress response—the hippocampus—is also vulnerable to cortisol. This creates yet another vicious brain circle: if the hippocampus can't function properly, it can't shut down the HPA axis, which leads to elevated cortisol levels, which leads to more hippocampal damage, which can lead to increased fearfulness, which leads to increased stress, which leads to increased cortisol production, which shrinks the hippocampus, which then can't shut down the HPA axis, and on and on and on. Repeated stress in rats causes increased levels of cortisol. So does depression, which will actually shrink the hippocampus. (Researchers are still investigating the long-term effects of this damage.) In rats, the dendrites in the PFC that are destroyed by cortisol can grow back, but there are also indications of permanent change.

Everyone knows stress is bad for you, but the science of chronic stress makes you want to drop out of American culture altogether and take up permanent residence in a warm bath. Over time, cortisol can destroy you. In a study being conducted at Rockefeller University, researchers are trying to determine if chronically elevated cortisol is responsible for the health problems in America's ghettos. There is an idea floating around out there that what we see as social problems or matters of education or class may also have a physiological component. That maybe the worst part of being poor is not what it does to your spirit, but what it does to the rest of you.

By the time I was in high school, Bill was in a state of chronic stress. It was the spring of 1985 and we were heading toward the Citibank ATM in Forest Hills. Bill and I were going to spend the day in the city, walking around, and maybe see a movie. It was a fine day: we were playing hooky; the sun was out; our windbreakers had epaulets.

I was happy to be out with him, but also wary. His zany side had taken on an aggressive edge of late. In our living room, after the nightly news, Bill would act out elaborate fantasies about making Ronald Reagan's head explode on national television, or possessing the president's body and making him dance and lisp like a sissy. He would come home with two hundred dollars' worth of classical music CDs the day after declaring the family was broke, or spend hours composing, and then reciting, a long list of what he thought were funny names for dildos. ("Nebuchadnezzar, lad! Nebuchadnezzar!") When my mom or I would protest, he would complain that he was only being "exuberant," as if this behavior were common the world over to the joyous, the blissful, and the happy-go-lucky.

On our way to the ATM that day, Bill spotted an old lady up ahead, approaching the locked ATM door. She was moving slowly, pulling out her wallet for the bank card, pausing for a moment with her shopping cart and supermarket bags. Bill broke into a trot down Forest Hills Boulevard, leaving me behind.

He beat her to the door. As she was fumbling with her card, Bill whisked by her, contorting his body to bend around hers, then popped in his bank card and opened the door just as she raised her head. He was already in, he had won, and she barely noticed (she was so old and slow and absorbed), but Bill took extraordinary satisfaction in his victory. I got the whole idea about needing to get ahead in New York in small ways, even if I didn't subscribe to it myself. But what I didn't get was why he had to laugh in her face.

Bill wrinkled his nose at her and shoved his face forward and all but stuck his tongue out at her. He made a guttural sound, a tortured *nnnnn*, that came from the back of his throat. I couldn't see her face, only the startle that stiffened her back. Then Bill disappeared into the bank and the old lady, thinking better of following this lunatic, walked off down the street. When I caught up with Bill, he was laughing, "cackling" as he called it.

"Outstanding!" he said. "Oh, lad! You should have seen her face."

"I don't know why you had to do that," I said. "There was no reason to do that."

"Nyah, nyah, nyah," said Bill. This was his response, the three dismissive notes he singsonged when someone criticized him. He was acting like I had been nagging him all week.

Bill then adopted his sissy voice. "Oh, I suppose I should have helped the poor little thing," he said. He stuck out his pinkies and pranced around. "The poor dear! The poor, poor dear!"

"Pitiful," I said. People were staring, but I had gotten so used to this kind of public behavior that I was incapable of being embarrassed.

"*Pitiable*," Bill said. "Pitiful means full of pity. Pitiable means an object of pity."

"You know what I meant," I said.

"No, lad," he said. "I didn't know what you meant, because you were using the wrong word."

"Okay," I said. "There is no need to yell."

"I'm not yelling!" He then turned his anger toward the line of people waiting to use the ATM.

"Slow," he said. "Inferior."

One of the other cruel tricks of the stress response is that cortisol enhances the fight-or-flight response. The more stressed you are, the more your amygdala is on a hair trigger, highly sensitive to both fear and aggression. The stress had Bill's cortical control out of balance. The Man of Reason couldn't be reasoned with anymore.

. . .

Between baby and brain I was feeling more nakedly emotional. Before the baby, I might have been anxious about the way President Bush was ruining the country, but now his policies felt like a personal attack on my family. The smallest pet peeves seemed like Horsemen of the Apocalypse. A part of me knew this was the cortisol talking—scrambling the inhibatory signals from my frontal lobes—but knowing about cortisol didn't make the stress go away. If anything, my being conscious of this unconscious regulatory system made me feel even more powerless against my body.

Fortunately, I had an idea, not so much for conquering my stress, but for finding its limits. If the fear-conditioning experiment proved that I couldn't wish away my stress response, then perhaps I would feel better if I went in the opposite direction. Rather than trying to prove that the cortisol wasn't a factor in my life, I planned to push my stress level as far as it would go, like an athlete probing the upper reaches of his endurance. If I knew the worst stress had to offer, then maybe I wouldn't feel so helpless in the face of the day-to-day variety.

Strangely, I still fancied myself something of an amateur neuroscientist. The lesson I had taken from the Mall of America was not *you will never, ever, ever be a scientist*, but rather *maybe you need to try again*. I went about setting up a stress lab in my house, scouring the scientific literature for ways that I might investigate this phenomenon, and entertaining fantasies of making an important discovery. This time, however, I was smart about it. Now when I thought about a breakthrough, I confined my imagined success to simply starting the scientific conversation. I would learn something profound and then the neuroscience community would put the finishing touches on my great truth.

As I started looking for an experimental protocol I could borrow, I wished I had more money. I might have even gotten me some animals. I once thought researchers studied rats, monkeys, and the occasional cat. I had no idea of the variety of creatures that played a role in modern neuroscience, nor the specificity of their attributes. There are neuroscience studies with sea slugs, songbirds, and lobsters. There were ovarectomized Syrian hamsters (a breed of hamster with its ovaries surgically removed) and dopamine D2 knock-out mice that were bred to lack that specific receptor. I don't even know how many types of rat there are, but researchers have their preferences. "I used to like Wistar rats," one scientist said to me, "but now I much prefer Sprague-Dawley."

Stress researchers who use animals also have more fun choices when it comes to experimental protocols. They have put sheep in isolation, withheld food from monkeys, separated rat pups from their mothers, and made prairie voles swim for uncomfortably long periods of time. The most common method for creating stress in animals is long-term restraint. Rats are put in soft, wire-mesh bags and held fast for up to six hours at a time—in some cases for as many as twenty-one consecutive days.

If I wanted to use humans, I had fewer choices, and I would also be constrained by medical ethics. One researcher told me about some pretty cool tricks to stress humans. Because the stress response is more robust when the stressor is outside the subject's control, British military researchers in the 1950s used to devise terrifying ordeals for their pilots in an attempt to determine the typical limits of pilot effectiveness. These ordeals made the research subjects feel like they were literally going to die. One experiment involved telling everyone on an airplane in flight that the plane was going to crash. Another had the subjects in the middle of a forest with a radio. Suddenly they were warned of a forest fire descending upon them, but the transmission "broke off" before they were told the direction of the fire, leaving the commanders scrambling around in the woods in an effort to save lives that weren't in danger.

Today the human stress experiments are more tame. In the Trier Social Stress Test, subjects are told that they will have to write and present an essay titled "What I Don't Like About My Body" in front of an audience, and that the unsavory event will also be videotaped. What they don't know is that the whole thing is a ruse to elicit a stress response, as preparation for giving them any number of cognitive tests. There is also the Matt Stress Response Protocol, part of which involves solving timed math problems while graduate students yell at you, and the Cold Pressor Test, which involves holding your arm submerged in ice water.

When I told Liz that I was going to do experiments at home, she got nervous.

"What are you going to do?" she said.

"Experiments," I said. "It's nothing. No blood."

I then asked her if she would wrap me in a bedsheet to mimic the restraint stress. I told her about the wire-mesh bags the rats get. "No way," she said.

"You only have to leave me alone for four or five hours," I said, noting that the rats sometimes got days.

"Dennis, I am not wrapping you up like some rat in a bag," she said.

And so I was compelled to continue my work in secret. I decided that it was best not to tell her what I was doing—not to tell anyone. I convinced myself that this was more real. True pioneers worked alone.

Late one night I was down in my office. Liz and Owen slept upstairs. My lab—as I called my office now—was in the back of the house. I didn't have to worry about disturbing them with my science.

I sat down to do a Trier, but because I wasn't "experimentally naïve"—in other words, because I knew I didn't have to give a presentation—I figured I would have to ratchet up the self-loathing in the essay about not liking my body. This, I thought—and I thought quite reasonably—would compensate for my awareness of how the experiment worked. I set the timer on my cell phone and started writing:

What I Don't Like About My Body
by Dennis Cass

I don't like my wrists. There is a spot above the joint moving into the forearm part. What is it, two inches wide? Three? I have these tiny forearms, too. I remember in college a friend said I looked like a skeleton, and the worst part about it was that he didn't say it to my face (I overheard him), and he didn't mean it as an insult. He was marveling. Or the other day when another friend said can you imagine what Owen will look like if he inherits my face. It was like she hoped Owen had any type of body but the kind of body I have.

I have been skinny for so long that I don't know if I remember how bad it used to be. The trip in junior high school, 7th grade, when I wouldn't take off my soaking-wet coat after the day at Six Flags because I didn't want the other kids to see my wrists. It's not even my wrists, it's the forearms. And to feel so afraid all the time, not even that you're going to get beat up, but that you don't have any leverage. You don't have any power when you're skinny. You carry no physical threat. No one is ever worried about pissing you off. People, and especially other men, will always relegate you because you lack physical size. So I spent a lot of time cowering and cringing and not even thinking of standing up for myself. Or crossing the street all the time. And now that I'm married and a dad and I'm invisible, I'm still waiting for someone to threaten me.

I don't like how hairy I am. I have very hairy arms and legs and my lower back has hair on it now. I don't like my receding hairline or how my face looks in pictures, like it's bent. I can't smile in pictures. I feel completely self-conscious, skinny on film.

How Bill used to tease me. He'll never grow up, he'll never grow up, he'll always remain a child, with wee, wee, wee . . . *He was so relentless. I remember just crying and crying and he was playing the piano even. And doing that weird, creepy witch stuff and why didn't my mom protect me? Why wouldn't she stop him from tormenting me and—*

The alarm on the phone rang: time's up. Which was okay because that essay was getting weird. I wasn't sure where that bit about Bill came from. I had been thinking off and on about Bill and his brain, but it had been almost academic. Until this essay I hadn't felt his presence, and yet here he was, starting to emerge. I pushed my laptop away and turned my attention back to my stress exercise. Bill had been shadowing me for months, but I wasn't ready to see the connection between us yet. I wasn't ready for him to be in the room.

Still, I was pleased. Hating myself on paper had certainly done the trick. I felt the downstream reaction of the stress response, the extended fight-or-flight. The response in the fear conditioning experiment was one kind of response—that was epinephrine. But I was becoming a kind of connoisseur of fear and stress: fear had a higher, thinner chemical feeling, whereas over the course of that essay, I had felt the dull, slow drag of cortisol. When I had gotten to the part about Bill I had felt the fight-or-flight response intensify again, but I dismissed that moment as what scientists call an "outlier," a bit of data that falls outside the norms of the rest of the data (and, hence, statistically not significant).

Then I wondered if my modified Trier was enough. It seemed that if I was going to find my personal threshold for cortisol production, I would need to push myself harder. So I threw together an impromptu Cold Pressor Test. I tiptoed into the kitchen and made every effort to be quiet as I filled a stainless-steel mixing bowl with ice cubes and cold packs from the freezer. Then I added a little water just to enhance the cold of the ice and took the plunge.

There is some debate over which is superior, the Trier-like instruments that stress the body indirectly or the Cold Pressor, which acts directly on the body. Proponents of the Trier say their test is more natural, mimicking the kinds of everyday stress that people actually experience, while the Cold Pressor camp point to their test's enhanced objectivity and repeatability. Like the shock in a fear-conditioning experiment, a bucket of ice is a bucket of ice.

At the moment, my skinny forearm submerged in wet ice, I had to give the nod to the Cold Pressor people. This was the real thing. My vision closed down, my body was in spasms. I could taste the cortisol working off my tongue. The ice was rattling around the metal bowl, and the freezing of my arm had me doing a dance that involved gripping the rim of the kitchen sink. I once read a story about how the meat from stressed-out pigs tastes bad, the cortisol causing a breakdown at the cellular level. I experienced the truth of that study firsthand. I had definitely become one unsavory piece of pork.

Then I went back into my office and set the timer again. Researchers typically measured blood plasma cortisol levels and salivary cortisol levels, the latter of which took about fifteen or twenty minutes to manifest itself.

"What are you *doing*?"

I turned and saw Liz standing sleepily in the kitchen. In some ways this was good: more stress. But I also would have liked to repeat this experiment at some point, and it seemed unreasonable, not to mention logistically difficult, to accidentally wake her each time.

"It's nothing," I said.

"What?" she said.

"It's nothing," I said.

There was a pause between us, one of those everyday standoffs between husband and wife.

"An experiment," I said.

"You woke me up," she said. "You woke up Owen."

"I'm sorry," I said. "Don't worry. It's done. All done."

She went back to bed. It was now 10:48 p.m. I was a good fifteen minutes past the Trier and the Cold Pressor and at the tail end of the encounter with Liz. The mental aspect of all three existed only as an echo now, the memory of what it felt like to write the body essay and wallow in ice. I no longer felt like bad meat—yet my cortisol levels were at their peak.

I had picked up a few home cortisol measuring kits from a company called Young Again Nutrients ("Male health and longevity" . . . "Life extension complex" . . . "Antioxidants anti-aging wonders"). I was still misguided enough to think that my lab was evolving. I was thinking that the problem with the Mall of America experiment was that I was so focused on the experience that I had neglected one of the core aspects of science: measuring things. Having grown as a scientist, I saw now that measuring was very important.

For $264 I was able to equip my lab with all the home cortisol tests I would likely ever need. I opened up a Young Again Nutrients cortisol kit, which contained two clear plastic tubes. I had already filled one tube this morning. Cortisol is naturally higher in the morning—it's cortisol that woke me up at 7:04 a.m.—and then tapers off over the course of the day. My hope was that by doing these experiments at night I would see an elevation that approached, or surpassed, the morning level. I took the lid off the tube and held it against my lower lip and drooled into it for the next few minutes. There was something satisfying about spitting into a tube and then sending that spit off to be analyzed. I could see that if the brain was ever going to be fully embraced by the world, this would be the way. You part with a little piece of yourself and a lab tells you about your mind.

The next day I sent off my kit to ZRT Labs, but I wasn't entirely satisfied. My encounter with Liz had pointed to a flaw in my research—it was too self-contained. I realized that the point of the Trier wasn't the writing of the essay and whatever self-loathing that process would generate but the threat of reading the essay in public. *That* was the real stressor. Stress is other people.

Animal research bore this point out. In one experiment, researchers took two tree shrews and put them in a cage together. The tree shrews, as tree shrews will, soon fought to establish a hierarchy of social dominance. After a period of separation, the animals were then reunited in the cage, only this time a wire-mesh wall kept them apart. What happened next will not please PETA. Both tree shrews showed elevated cortisol levels, but the subordinate one had the worst of it, suffering hair loss, weight loss, and an increased susceptibility to illness. Having a social relationship but being unable to act on that social relationship turned out to be stressful for both. But for the subordinate one it was even worse: the wire-mesh wall allowed him to experience the threat of the top shrew but never allowed the threat to complete itself.

Maybe this was Bill's problem with New York. It wasn't the disappointment of the city not being what he dreamed it would be. The agony for Bill was that New York was indeed a giant salon, but that he was stuck on Wall Street, or crammed in the subway with the ignorant masses, or relegated to Queens. He knew where the party was, but he wasn't invited, and merely living in the city wasn't enough. Maybe this is even the larger problem with New York in general. The city is at once transparent and inaccessible; there is always a piece of wire mesh separating you from who you want to be.

In this light, I realized that one of the biggest stressors in my life came from professional jealousy. The stress I felt as a result of my marriage or my friendships felt routine compared with the toxic stress resulting from the imaginary relationships I had with celebrities, media figures, and other writers—in particular Malcolm Gladwell. I read his pieces with a poisonous mix of envy and rebuke, appreciating his complete mastery of idea and telling example, and feeling falsely superior to him for his not being sufficiently messy and human. I have debated him on morning talk shows; in my head I've written lengthy takedown pieces—in one scenario it actually runs in *The New Yorker*, where Gladwell is surely Employee of the Month. He is my dominant tree shrew, and the wire mesh separating us is that I know who he is, and he doesn't know who I am, and if he did, it wouldn't matter because he's a better writer on a bad day than I'll ever be on my best.

A couple of days later I tried another experiment. At the same time of night as I did the Trier (again: consistency, control), I spent an hour warming up by reading conservative pundit blogs, and looking at all the brain books on Amazon.com that people were probably going to enjoy more than mine. Then I went over to Gladwell's website. Even the loading page made me feel like a loser. The clean, tight design, the effortless organization, and then, of course, the trophy wall: one perfect *New Yorker* piece after another, best-selling books, watertight writing. I felt what autoworkers felt when they lost their jobs to robots: you can hate the machine, but you can't deny that it's a superior welder.

I felt incredibly stressed as I was doing this. The subjective response was much greater than in the Trier, almost to the point of panic. I get this feeling while reading Gladwell that everyone compares my writing to his, side by side, my mistakes magnified by his perfection. I was going to fail and I was going to fail big and it was going to ruin my family. Surely this would be the biggest cortisol reading of all.

But it wasn't. My first report from ZRT Labs was the control. I had taken a reading on a normal day of working, helping take care of my baby, and keeping myself entertained as best I could. Without any extra experimental stressors, my morning cortisol level was 7.4 nanograms per milliliter, my evening reading was 0.4 ng/ml. ZRT had provided a range of what they deem acceptable. (Since cortisol was naturally higher in the morning, there were two ranges.) Going by their measure, my morning cortisol was at the high end of normal (3–8 ng/ml), while my nighttime cortisol was just shy of the low end of normal (0.5–1.5 ng/ml). The Trier plus Cold Pressor plus Liz came in at 1.7 ng/ml, but that was still only 0.2 above the top range. The professional-jealousy experiment netted me a paltry 0.9 ng/ml.

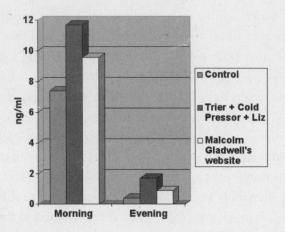

I was stunned. The professional-jealousy trial, which had been the most subjectively miserable, turned out to mean nothing to my body. It was as if the lab had said, "Quit whining."

Even more disturbing was my morning cortisol: 9.6 the day of the Trier, a hefty 11.7 the day of the professional-jealousy experiment. None of these levels reflected any additional pressure I was putting on myself. Hard as I was trying to stress myself out, ordinary life was proving to be far more stressful than anything even my runaway imagination could dream up.

THE SPLIT BRAIN

In the summer of 2004 I reached the point in my research where I was farthest away from my original intent. Rather than trying to figure out how my wits worked—a plan I had all but forgotten—I was now struggling to prove myself capable of even a single scientific thought. My failure at the Mall of America and the shuttering of my stress lab had only made me more desperate to prove that I wasn't a fool. Abandoning data collection, I switched to pure theory. I didn't care if this was a disease of the will, I would redeem myself with the Big Idea.

Drawn back to the prehistoric brain, I recast myself as an amateur evolutionary psychologist. I became interested in the notion that the structures of the brain that benefited our ancestors had now been hijacked by the postmodern world. I became a kind of brain-conspiracy theorist, bending the news of the day—war in Iraq, dirty election tricks, blogs—to fit my hypothesis that the species was tragically off track. It seemed to me that the media, both liberal and conservative, were engaging, perhaps unwittingly, in the systematic manipulation of the nation's collective amygdala. Pundits on both sides were shutting down our PFCs until no one could think straight.

I then extended this idea to anyone who worked in media and entertainment. Smart enough to know that I didn't have the intellectual chops to take on the neurobiology of Fox News, I focused on something that was more in my region of proximal learning: celebrities. I reasoned that famous people were performing the same trick, jamming our brains with emotional stimuli until we forgot who exactly we were.

One day, while watching VH1, my theory was confirmed. While watching *Best Week Ever*, I marvelled at the love-hate relationship the show's comedians/pundits had with celebrities, lauding them with one hand and slapping them down with the other. I went back to my evolutionary-psychology papers and found a line in Cosmides and Tooby's primer that had the answer. "It is easier for us to deal with small, hunter-gatherer-band–sized groups of people than with crowds of thousands," they write. Yes. We have this love-hate relationship with celebrities because they are—through some freak of development, or genetics—masters at communicating with mass audiences. We are fascinated by how they are able to use fame to exert power and control. Yet we hate them because these same abilities mock our more rudimentary survival skills.

This idea felt right. Unlike, say, Small Hole Theory, this was going to be the one.

I felt it in my gut, and scientifically speaking, trusting your gut is not a bad idea. While the brain hogs the spotlight, it's only part of the central nervous system, whose visceral nerves transmit and receive information from all over the body. According to Antonio Damasio, formerly head of Neurology at the University of Iowa and now director of the Brain and Creativity Institute at the University of California, these nerves not only register the subjective experience of pleasure and pain, but they also help the brain use this information for higher functions.

Damasio calls this theory the "somatic-marker hypothesis." He maintains that sensory information from the body (*soma*) gives the brain physical context for rational thought. In other words, there is no such thing as a coldly rational decision.

In one test of this idea, Damasio and fellow University of Iowa researcher Antoine Bechara invited patients with brain damage to gamble. In the Iowa Gambling Task, subjects and controls are given a supply of imaginary money. They then choose from four decks of cards on a computer screen. Two of the decks are "good decks" that pay out smaller amounts but also result in fewer and smaller losses. The other two are "bad decks" that pay out big, but they also cause catastrophic losses. Picking from the bad decks eventually bankrupts the player.

Damasio and Bechara discovered that the brain structure at work during the Iowa Gambling Task is the orbitofrontal cortex, which helps you match the visceral (or gut) feeling of winning or losing with traditionally rational observations about whether those wins are coming from good or bad decks.

I have done gambling tasks such as this one, and it is painfully obvious what you need to do to win. It's not even a conscious decision, or even an aha! moment: after a number of draws you can tell which decks are the winners. But patients who have orbitofrontal damage have trouble making the connection. It's as if their gut feeling about what is bad or good has been removed. Intellectually they can process the *idea* of a bad deck, but they can't *feel the badness*, so they continue to draw cards that empty their bankroll.

British researcher F. C. Murphy has found that subjects experiencing a manic episode behave similarly to patients with orbitofrontal damage. In his own gambling tests, Murphy found that "manic patients, as a group, displayed a heightened tendency to choose the less likely of two possible outcomes, attempting to earn reward on the basis of the less favorable response option." Even when the odds were a daunting ten to one, the manic patients gravitated toward the wrong decision.

Now I saw that when Bill really started to fall apart he didn't necessarily know any better. When he would make bad decisions—buying classical music CDs when he should have gone grocery shopping—I had always thought that this disregard for the family's basic needs was a choice that he had weighed in his mind. He chose the expensive steaks over the staples because he valued his own pleasure over the good of the family. He was capable of doing what was right, but declined out of stubbornness, selfishness, and greed. It wasn't until he really started to lose it that it became clear that there was something else at work.

One night, when I was a junior in high school, I was awakened by a sound coming from the front of the apartment. I got out of bed with a pretty good guess at the source. The light was on in the dining room; on my way, in the kitchen, the telltale signs that Bill had been up for a while: a collection of knives with marshmallow topping and devil's food crumbs stuck to them, plus a half-gallon of Breyer's Fudge Ripple ice cream melting in the sink.

"Lad!" Bill said as I approached the dining room. "I'm glad you're here. There's something I want to show you."

"What are you doing up?"

"Come see."

I entered the dining room, the place where all the heavy, unused things went. There was the piano Bill never played; the sewing machine, an elaborate, indulgent gift Bill bought my mom one Christmas, which she rarely used; Bill's ponderous vinyl record collection he never listened to; and the stereo I was not supposed to touch. No one ever used the dining room except Bill during tax time, or when he was on deadline for a municipal bond review, or on nights like tonight, when he was working on one of his projects.

Bill had out pencils and a heavy silver ruler that my mom had brought home from the ad agency where she worked. There was graph paper covering the dining room table and more pages crumpled at his feet. Wearing shorts and a T-shirt, Bill was bent over the table because he was too wired to sit down. Lately he had started having these physical tics—a flicking of his heel to his ass, as if his thigh muscles needed to have extra energy worked out of them.

"You should be in bed," I said. "You have work tomorrow."

"In a minute. Come here."

I walked around to see what he was working on. It was a floor plan of a house. There were double lines for walls. Swinging arcs for doors. I have no idea where he learned the language of draftsmanship, but this looked like a regular blueprint. The rooms were small and packed tight. There were two stories, each on its own set of graph paper. It was a house, but who knew where, or, more specifically, when.

Bill seemed to be slipping in time. The more he explored New York and learned about it, the more he started letting on that he had been born in the wrong class, the wrong country, the wrong age. He started talking about how wonderful it must have been to have been landed gentry in England, or perhaps one of the early barons of New York. He would hop in a time machine in a second to be back in the glory days of old England, where they would instantly recognize him as a man of worth. Bill liked to hang around Sutton Place, where he talked about the brownstones with a kind of proprietary intimacy; should an owner spy him through the window he'd surely invite Bill in for a glass of port.

But now the fantasy was stretching out and asking for more space. Bill had taken an interest in Victorian architecture and he had been up all night designing a single-family dwelling. The weird thing was that when Bill started on his description of the house, it was as if this little place near London wasn't where we might have lived if it were still 1875, but where we were going to live tomorrow.

"Okay, what we have here is the entryway where people will, of course, enter the home," he said. "It's customary for guests to be shown into the drawing room. Of course there will be servants—a cook, a butler, and maid at the very least. One of the servants will offer our guests a drink, while the man and lady of the house ready themselves."

"That's great, Bill, but you really should go to bed. You have to go to work tomorrow and—"

"Wait, wait, wait," he said. He flourished the pencil and gave his heel a kick up at his ass.

"I really want to go to bed, and you should, too," I said. I knew that Bill was having trouble at work. He had already left S&P because he couldn't get along with people and had now moved to Citibank, where everyone thought he was so weird that he occupied a cube on a different floor from his group.

"C'mon, I just got started," he said, like he was trying to be pals. "Please. Lad. Indulge me."

I wasn't old enough to be articulate about my concern over his career, but I was aware enough to know that he was capable of losing his job if he didn't get to bed, get some sleep, and stop fucking around. So I folded my arms to make sure that he saw my protest toward this inconvenience and lack of respect for my interests and need for sleep, not to mention his reckless behavior—staying up late when he needed to be fresh in order to protect his job. He would have my attention, but he would pay the price in heavy sulking.

He ignored me. "Okay, here in our conservatory we'll have our *pianoforte*," he said. "We'll wile away the hours playing music or listening to a recital. In Victorian England people still relied on each other for entertainment, so it wasn't uncommon for even a middle-class family to host recitals in their home. Mendelssohn isn't going to come over, but you make do, right?"

Here Bill allowed himself a laugh, and then pointed to where he had drawn in the piano. Bill's freehand drawing wasn't very good, so it was hard to make out, plus the page was grayed and smudged with multiple erasings. Looking around, I saw the earlier drafts. There were some that seemed perfectly fine, but for whatever reason, Bill had started all over again. Some pieces of paper just had a line on them, but the line was a little off, the charcoal nudging away from the pale-blue graph-paper lines almost imperceptibly.

"And of course, there is the library. Can you imagine? A library! Imagine having an entire room just to *read* and, and, to *think*." He circled his pen around his mind, as if drawing up a soft-serve ice cream cone of enlightenment. I let out a theatrical sigh in the hopes that he would notice the absurdity of this moment, but he didn't see. He was going to keep us both up all night, and it wasn't because he didn't care about the consequences. It was because he couldn't feel them.

I tried to employ my own version of the Socratic method— which he had insisted I learn—and built up the argument.

"You agree that you have work tomorrow," I said.

"Of course."

"And you agree that it's late and you need sleep in order to function at work tomorrow, and that work is important," I said.

"Granted." Bill was smiling now. He saw what I was doing and was amused that I was advancing an argument. He looked like he was about to comment but let it go.

"Then you would also agree that in order to get the sleep required to function at work tomorrow in an effort to survive that you should go to bed," I said.

"But I want to finish the conservatory," he said.

Then we reached the point in our night when I saw that it was never going to end. I announced I was going to bed. This time Bill let out an even more theatrical sigh. "Very well," he said. "But you know. And I'm just being honest. But sometimes you're just not very much fun." From the tone of his voice he was making it clear that he wasn't disappointed for himself, for not having an audience. Nor was he disappointed *in* me. He was disappointed *for* me, the disappointment I should feel for myself were I not too blind to see how un-fun I was.

"All right," I said. "Let's stay up all night. It will be so much fun, Bill. Let's just be fun, fun, fun."

"Excellent!" he said. "Spicy!"

Perhaps the defining characteristic of mania isn't "manic energy" but rather a string of bad decisions that gives the illusion of being, as Bill would say, "jazzed up." It's similar to looking at anxiety as either an overactive amygdala or an underactive prefrontal cortex. Bill's zany behavior and his exuberance were not a surplus of fun, but an absence of being able to see the consequences of his bad decisions.

What was my excuse?

Despite the obvious failures of my previous experiments, I had yet to be chastened. As summer turned to fall, I was well into my evolutionary-psychology efforts. I picked Bill Maher as the celebrity who best embodied the complicated relationship between the famous and the consumers of fame, at least for me. Maher was someone I both loved and hated—appreciating his politics and his willingness to speak truth to power, but also resentful of his glibness and his cockiness. Even though I often agreed with him, I felt his methods were just as destructive as those of his conservative counterparts. He was shortchanging his fans, denying them the ability to think for themselves by leaving no room for more moderate points of view. From an evolutionary-psychology standpoint he was using his personal magnetism to exert subtle reproductive pressures on his audience, or tribe. If I could only partner with another scientist, maybe he or she could help me solidify my findings.

I contacted Paul Whalen at the University of Wisconsin. Whalen's lab was investigating the role of the amygdala in terms of its role in anxiety disorders, but more importantly he worked with faces. In the sixties, researcher Paul Ekman photographed actors portraying universal expressions of fear, anger, sadness, happiness, surprise, and disgust. Whalen and others now used these Ekman faces as stimuli to gauge how the brain—specifically the amygdala—reacted to human emotion. A reputable and respected researcher, Whalen would never overreach like I was, which is probably why when I set up my appointment with him I left out the fact that I was going to ask him if he thought Bill Maher could induce a personal, not to mention culture-wide, amygdaloidal reaction.

Before my trip I started gathering evidence. Up at night in my office—no longer a lab, but now more of a one-man think tank—I surfed the Internet, looking for pictures of Bill Maher. It was important to find one that summed up his Bill Maher–ness. I found: a shot of him smirking at the camera against a plain backdrop; Bill Maher at what appeared to be the Genesis Awards, sponsored by the Ark Trust and Animal Planet; a screen grab of him on what looked like the set of *Politically Incorrect*, giving the guests a kind of open-arms, can-we-be-real, c'mon-people-stop-being-idiots gesture; Bill Maher smile-smirking in front of an American flag; Bill Maher in a black turtleneck, leaning on his folded arms and smile-smirking, but a little sad, as if to show vulnerability.

On the trip to Madison I was excited about my meeting with Whalen—young neuroscientist meets young amateur evolutionary psychologist. It was a beautiful September day, perfect for a drive, perfect for science. We were going to get something started here, and I could picture Whalen's reaction, as if I had served him up exactly what he had been searching for.

I met with Whalen in his office in Madison (he has since taken a position at Dartmouth). He was as I had imagined: around my age, a bit of a hipster, with an intelligence that he displayed casually. Outside his office I was pleased to see a Kerry-Edwards sticker and a joke candy wrapper that read AMYGDALA JOY. I also liked Whalen because he was a little anxious. Joined by Ashly McLean, one of his graduate students, we sat in his office, where Whalen twirled an empty water bottle in his hands. Later he would switch to a thumbtack.

After some introductory talk, I busted out the goods. "I brought some stimuli of my own, if you don't mind," I said. Reaching into my bag, I produced my Bill Maher printouts.

Whalen grunted amiably. "Bill Maher," he said. "What about him?"

Yes, Dennis, what about him? If I had any concerns about having orbitofrontal damage, I could stow them away, because I suddenly felt the badness of my decision in visceral nerves all over my body. Everything I had ever thought about evolutionary psychology and the amygdala and the neurobiology of fame soured on the spot. Immediately I felt the downstream effects of an amygdala reaction—the sweaty palms and the shortness of breath—only this was in the service of embarrassment. I should have just put them away and made some small talk about the upcoming election, but instead I kept going.

"When I see him," I said, "I feel like . . . I feel like I get . . . a fight-or-flight."

"You're afraid of Bill Maher?" said Whalen. The tone of his voice was appropriately skeptical.

"I wouldn't say I'm *afraid* of Bill Maher," I said.

"Are you suggesting there is something about this face that sets you off?" Whalen said.

I was all flustered now. "I was wondering if just as an exercise we could talk about this face. Because you do faces. What can we say about this face?"

I was referring, of course, to the Ekman faces. The faces Whalen used for his famous experiment where subjects saw subliminally angry faces amidst the neutral faces. (Whalen even had the Ekman face for anger posted on his wall—the actor looked like a pissed-off android.) I picked a photo that showed Bill Maher looking directly at the camera. There was silence as

Whalen and McLean looked at the tiny, grainy image. Whalen had taken my suggestion that we do Maher's face literally. Whalen's work with faces and emotion didn't rely on any meaning other than the raw data provided by the expression. There was no cultural context, only pure emotional value.

"It has zero percent emotion," Whalen said. "It looks like a twenty-five percent smile. . . . It's asymmetrical . . . a half-smirk."

They were both looking, both he and McLean, or at least making a show of looking. I was getting increasingly embarrassed. This wasn't science. This was barely even journalism.

"Literally," I said, "when I see his face, I get, like—"

"Who is he?" McLean asked. At first I thought she was kidding, but no. Her ignorance of him, though a small victory, was appreciated. She was, at that moment, anyway, my favorite person in the whole world.

Whalen explained who Bill Maher was in a way—praise for his intelligence and political commentary, awareness of his show business résumé—that made me think that if not a fan, then Whalen was at least a sympathizer. I felt a new shading to my already prodigious embarrassment—*He probably thinks I'm a Republican.*

"He's really hard on his guests," Whalen said. "He does seem like he would probably be a dick in real life."

This statement emboldened me. Maybe I was on the right track.

"To me," I said, "this is my snake."

"Wow," said Whalen. He was a little taken aback. For me, I was simply throwing out words, but for Whalen the snake was the lay shorthand neuroscientists used when they wanted to express the full activation of amygdaloidal complex.

"Does he look like someone who wronged you?" said Whalen.

If he had said, "Does Bill Maher remind you of your step-father?" I might have allowed myself the connection. Instead, like Bill during his intake with Dr. M., I made a show of thinking about it. The smug certainty. Cutting people off. Having the last word. The raw intelligence and sharp wit that made the asshole-ish-ness harder to dispute, or at least easier to excuse. The strong point of view, the pugnaciousness, the inappropriate humor, the inflated sense of self-importance, the creation of the hermetically sealed world in which all serious challenges are eliminated.

I said, "I don't think so."

"Maybe he said something you didn't like," said McLean.

"Do you think it's something about him that you don't like, or is it the features?" said Whalen. "Do you think he's skeezy in real life?"

They were really trying to help now, which was great, because I was starting to lose it. It was so clear that I had no business doing this, and it was painfully obvious to everyone in the room, and I started to talk fast and sweat and shake.

"You get an actual visceral response?" Whalen said. He still wasn't sure what to make of this.

"The reason why I brought this . . ." I said. "It's an example of all the stimuli that are not relevant—like, my cortex says, 'Who cares?' But there is something in the emotional brain that says, 'Change the channel now.' It's like there is a conflict. Between the emotional brain and thinking brain."

Finally Whalen got what I was trying to say, even if Bill Maher couldn't really be my snake. Nothing could be my snake, except in the realm of exaggeration.

"The answer for you might be there," he said. "But that same feeling can be driven by some very complex systems. The more complex the stimulus that's setting you off, the less likely the amygdala is pulling this off on its own. But that said, the more experience you have with that idiosyncratic thing that sets you off, the more likely it is that the memory could be stored in your amygdala. But I would stick with the first one. I don't know. Whether you know it or not, he reminds you of somebody bad, or things that are bad. Maybe there was another person in your life who was really a nasty person and they have some quality in common with him, either the way they smirk or the way they look.

"Our definition of 'threat' has expanded. Now, there is threat to our ego, to our feelings. Why wouldn't you use this amazingly efficient, highly connected system to protect you from that, too? If the amygdala's original job was to keep you from getting physically hurt, then maybe its job in the present day is to protect you from psychological injury."

"Do I have a Bill Maher neuron cluster in my amygdala?" I asked.

"You mean, is there a Bill Maher assembly?" he said. "I doubt it. I think it's too highly organized a stimulus. It has to be a network. But . . . the amygdala could have been modified. That might have happened."

"I've got *Real Time* on my TiVo," I said. "Maybe I just need to face it."

"Do some desensitization," Whalen said. "Start with the chair twenty feet away. Then fifteen feet, then ten, then five. Hopefully he's funny that day."

• • •

The brain is tricky, tricky. If one part of the brain isn't working correctly, the effects can ripple through the whole organism. Take the case of what is called a "split-brain" patient. In an effort to alleviate epilepsy, neurosurgeons sometimes sever the corpus collosum, a thick bundle of nerves that acts as a mediator between the two hemispheres.

Split-brain patients act normally, but when you present stimuli to only one hemisphere you get interesting results. It's as if the split creates a kind of very specialized blindness. "People become paranoid and think people are hiding stuff from them," one researcher said to me. "If they can't see the fork they say, 'You took my fork.'" Present a dirty word to their right hemisphere and they will become embarrassed, but won't know why. The right hemisphere gets the meaning, but without the language component from the left the source of the embarrassment is a mystery. Conversely, show a picture of a relative to the left hemisphere and the split-brain patient may not recognize him or her.

In an experiment by Dartmouth's David J. Turk, a split-brain subject is shown a series of pictures of himself that have been digitally mixed with pictures of researcher Michael Gazzaniga, author of *The Ethical Brain*. On one side of the spectrum are pictures that are 90 percent research subject, while at the other end of the spectrum it's almost pure Gazzaniga. (Gazzaniga is familiar to the subject, so the latter's brain isn't parsing a stranger—these are both faces the subject knows.) In between, the features of the two are mixed together, morphed into a composite to varying degrees. The subject is then presented with what seems a simple question: Is it me or is it Mike?

For the subject, the bias of his left hemisphere is to err on the side of thinking the morphed images are himself, while the bias of the right hemisphere is to think it's Gazzaniga. Turk writes in his paper "Mike or Me?: Self-recognition in a Split-brain Patient" that "[both] hemispheres were capable of face recognition, but the left hemisphere showed a recognition bias for self and the right hemisphere a bias for familiar others. These findings suggest a possible dissociation between self-recognition and more generalized face processing within the human brain."

Why is this important? Turk notes that "to operate effectively in the world, people must be able to distinguish between 'me' and 'not me.'" Without the split-brain patient study I would take this for granted, but now I was grateful for the distinction. To be myself I had to first not be other people. Part of being a whole person, then, is perhaps some kind of back-and-forth between the two, a constant checking between what is you and not you, what is true and what is false, but I had been so caught up in the quest to see how my wits worked now that I was neglecting my past.

Whalen and I parted company; he seemed to wish me well. Afterward I stopped at Culver's, a midwestern fast-food chain that cooks burgers in butter. The woman at the counter was excited to hear that I had never eaten a butter burger. I found the experience depressing. Researchers like Damasio made pilgrimages to The Hague and Spinozahuis and I was at a Culver's, trying to figure out why this entire endeavor was such a mess. I watched people eat their butter burgers and gulp down frozen custard, and it made me think about something my stepfather used to say. *The ignorant masses.*

I thought this was funny. I hadn't thought of that phrase in a long time, so separated was I now from Bill, how far apart we were in term of miles and personal closeness. But this little phrase was enough to give my subconscious something to chew on. Later I was in my office sifting through my notes and trying to piece my brain together when I started making what to me were unexpected connections. I started to see strange parallels: the hubris of the early days and the blindness to my own abilities when taking on the brain; the inability to recognize how much trouble I was in at the State Fair and in Lehericy's office; the excess of anxiety and overreacting during my fear-conditioning experiment; the misguided anger at popular science; the further denial of what my stress research was doing to me; the grandiosity of thinking I could be an amateur neuroscientist, not to mention the absurdity of trying to run my own lab; and then finally the stubborn denial of my own foolishness, even when it was plainly presented. I had yet another insight, nice and warm in intensity but ice-cold in its inescapability.

I'm turning into my stepfather, I thought. *And the brain is my New York.*

WHAT BRAINS LIKE

I spent the next month cringing. In my head I kept reliving the moment when I reached into my shoulder bag and pulled out those printouts of Bill Maher. Like a car explosion in a low-budget action movie, it played again and again from multiple angles. All thirty-six thousand members of the Society of Neuroscience knew I was a complete ass (Nowture, anyone?). I would be walking around my neighborhood and suddenly think about an exchange with a neuroscientist and shudder at what a fool I had made out of myself.

Scientist #1: "Did you hear about that writer who came in here the other day acting like there was some kind of war in your head between the amygdala and the prefrontal cortex?"

Scientist #2 (rolling eyes): "I can't even talk about it. I feel too sorry for the guy."

If I was professionally embarrassed, the thought that I was turning into my stepfather made me sick to my stomach. Ever since my junior year in college I had defined myself as being the opposite of Bill. I had taken the melodramatic teenage I'll-never-be-like-you speech and turned it into a personal mandate. Living in Minneapolis, trying not to judge people too harshly, checking myself for not being grandiose or overly ambitious, respecting other people's points of view—I had thought I had succeeded in becoming an anti-Bill of sorts. I even used to fight with him over the phone, arguing with him over his outlook on life, while he would express disappointment in me for turning soft. "I can't believe you actually live in Minneapolis," he would say. "It's such a non-city."

Whereas before I could keep some intellectual distance between the two of us, now Bill had an almost physical presence—Bill rode with me everywhere I went. He had hijacked my project, and if I was indeed turning into him, then that meant looking into topics—depression, addiction, mania—that I would have preferred to leave alone, and reliving times from my life that I preferred to forget.

Sometime early in my senior year in high school Bill stopped by my room. He took a seat on my bed, where I had been reading. We weren't talking much at this point. It was the fall of 1985 and I was about to start my tour of potential colleges. I couldn't wait to get away.

"Lad," he said. The tone was serious. "I have to tell you something."

Bill was sitting right next to me on the bed, something he never did. In our living room there was a love seat, a leather chair, and a couch; we each had our stations. We rarely talked about anything personal. It was all books and movies and TV. Bill's earnest tone put me on guard. Anytime Bill became fatherly it usually meant something bad had happened.

"Your mother and I," he said and then paused, his eyes up at the ceiling in thought. He made an effort to be courtly.

"I don't want to alarm you," he said and stopped.

"I don't want to call it a problem, because it's not a problem anymore, but your mother and I have had some problems with prescription drugs—it's nothing really."

Bill then spilled out the story of their addiction. For the past ten years they had abused prescription painkillers like Percocet, Percodan, Valium, and Demerol. It sounded serious, only Bill didn't use the word "addiction." Their drug history was less about dependence or abuse and more about how pills were once a lot of fun but had become "kind of a drag."

"Not that it's physically addictive," Bill said. "It's really more psychologically habit forming. You know, you do something for a while and it's fun, and then you don't want to do it anymore."

I asked him how it all started and he said it dated back to when we were in Seattle.

"You know, it's actually kind of funny," he said. "We would be at a friend's house and your mother would be on the lookout while I was upstairs frantically—and I mean *frantically*—rummaging through the medicine cabinet looking for Valium!" He let out a sigh, the kind you have after a big laugh.

He regaled me with tales of how they would con doctors or find ones who were easy with "scrips," and about fake trips to the emergency room, complaining of phantom migraines. The way he described it, the whole escapade was mostly comic, and even perhaps the fault of a medical establishment that wasn't as careful as it should be. But they were on methadone now, and they were fixing the problem themselves. He scoffed at support groups and didn't once mention treatment. There was no "Help me, Jesus" or a sense of personal failure or weakness or even struggle. They had it under control.

"We thought now that you're going off to college you should know," he said, then quickly added, "and it's not like we're heroin addicts, though you'd think how some people treat us we *were.*

We go down to this clinic near Wall Street, and there are real—I mean, whoa. There are some—whoa—some pretty rough characters. I don't know why they don't have a better way for us to get the stuff, you know. It's absurd. We're professionals. That we have to consort with drug addicts, with their *gimme*, *gimme*, like we're abusing street drugs."

"That explains a lot," I said.

Bill asked me what I meant and it all came out at once: the weird hours, strange attitudes, the rages, the nodding off on the couch, his habit of slapping his own face, which was something Bill did all the time, like he was interrogating himself, when in fact he was just trying to stay conscious against his body's better judgment, the fact that they weren't taking care of themselves, and that I had to do all the family's shopping and take care of the day-to-day operations of the house, like doing laundry, and how it always felt like we were on the verge of ruin.

Bill looked at me with pride.

"That is very perceptive, lad," he said. "I'm really impressed. You picked up on all that?"

We went out into the living room. My mom was on the couch, reading a Trollope novel. She was pregnant and due in December. She wore the same kind of mischievous grin that she had when we had talked about moving to New York, the shoulders up at her ears, like she had been naughty. Bill heralded our arrival with the fact that I had been so perceptive about the apparently not-very-well-hidden behavioral aspects of their addiction.

"You're just so perceptive, lad," he said.

My mother agreed. There was no apology. No real remorse. No accounting of how their addiction had made my life weird or painful, and the strange thing is that I didn't really care. Because they thought I was bright and perceptive, and I was just a sucker for that. They reassured me it was no big deal and that everything was going to be fine. They had come clean.

Since they came clean I felt I needed to make a confession of my own. They knew about my drinking because I was allowed to have a beer in front of them. I had drunk on vacation with them when I was as young as fifteen, getting goofy on Coors at my cousins' condo in Florida.

"You guys should probably know that I've smoked a little pot," I said.

They said it was okay. There was more pride, this time at my forthrightness and candor. After all, they had just shown me that there was nothing wrong with drugs as long as you dealt with them honestly.

I thought of their drug problem almost entirely in terms of feeling good or not feeling good. I knew about drug addicts, but at the time, I saw everything through the lens of the New York crack scare. The stories going around New York were of people instantly addicted from one hit, and of lives almost as instantly destroyed. Something like prescription drugs, which I associated primarily with Elvis Presley, didn't seem the same. The idea that a drug could be psychologically, as opposed to physically, addicting made sense to me.

Today the science of addiction still isn't entirely understood, but the underpinning brain functions of motivation and reward enjoy a little more clarity. It seems like it should be easy: the prehistoric brain seeks food, water, and reproductive advantage. You feel hunger or thirst or the desire to mate and you take the appropriate action. You itch, you scratch.

Sometimes you scratch hard. In a famous experiment conducted in the fifties, rats, with no further encouragement, would press a lever more than six thousand times per hour to receive a short pulse of electrical stimulation directly to the brain. Another study from around that time showed that animals would starve themselves when offered a choice between food and water and direct electrical stimulation. And, if electricity isn't your thing, there are a lot of other things in the world that satisfy more indirectly. In reward experiments spanning the decades, lab animals and human research subjects have been presented with juice, money, grapes, cocaine, chocolate, photographs of food, and *Bizarro* cartoons.

The early research into reward focused on satisfying natural appetites. In the sixties a scientist named Vincent Dethier discovered two eating reflexes in the fly. There was an "excitatory reflex," which made the fly eat whenever it landed on food, and an "inhibitory reflex," which signaled for the fly to stop eating. When Dethier cut its sensory nerve, the fly would stuff its fly face until its teeny-tiny fly stomach exploded. Findings like these led scientists to believe that reward functioned similarly to stress. The body sought homeostasis, a balance between hungry and full, thirsty and quenched, horny and spent.

But it's more complicated than that. If reward were some kind of internal homeostatic goal, then why do dogs that are given their full nutritional requirements still eat normal meals by mouth? Or why would a male rat work through a maze to gain access to a sexually willing female even if he was repeatedly taken away before he could consummate their love? There is also a wonderful anecdote about a research subject called "Tom." In a freak childhood accident he sealed shut his esophagus while eating hot soup. As an adult, Tom fed his stomach directly through a surgically created hole in his midsection, and yet still chewed his food. If he didn't put the food in his mouth first, he complained of not feeling satisfied.

This difference between what the body requires and what it desires drives one of the dominant theories about reward, the "incentive salience model." The incentive salience model divides reward into two separate but interrelated brain mechanisms: liking and wanting.

In his review paper "Motivation Concepts in Behavioral Neuroscience," Kent Berridge, a psychologist at the University of Michigan, says that liking is "essentially hedonic impact—the brain reaction underlying sensory pleasure—triggered by the immediate receipt of reward, such as sweet taste." In one study, researchers give, alternately, citric acid and sugar to rats, monkeys, and newborn human babies who have yet to be fed by either breast or bottle—the first thing they taste in their life is Science. The reactions to these substances are the same across species. The citric acid causes baby mouth gapes and downward tongue protrusions, as well as head shaking, arm flailing, and "aversive scrinching" of the nose, while rats do "gapes, head-shakes, face washes, and paw flails." For the sweet, human babies stick their tongues out and up. Their faces relax and they smack their lips and lick fingers, while rats exhibit "paw licking, lateral tongue protrusion, rhythmic midline tongue protrusions." What's remarkable is that the tongue protrusions are scaled. A gorilla may lick his lips slowly, a rat quickly, and a human in between, but they are all governed by the same formula: duration in MSEC = .26 (species' adult weight in kg)$^{.32}$. When it comes to sugar, we are all the same animal.

We were starting to see this kind of liking in Owen. Even at six months he had *preferences*. Certain strained foods—sweet potatoes, lamb dinner—went down particularly well, while others were less popular. He showed other, non-food, preferences, as well. When it came to his Lamaze plush aquarium, he definitely liked the little crab that went *crinkle, crinkle* more than he liked the little fish that went *jingle, jingle*. As for the magnitude of liking, there was nothing that elicited pleasure more than a television show called *Boohbah*. Even though the American Academy of Pediatrics warned against showing children under two any television at all, *Boohbah* was so charming with its rainbow of plush, jelly-like characters—not to mention so entirely mesmerizing for both Owen and myself—that we risked the long-term damage once a week for one half-hour, maximum.

But Owen didn't "want" yet. His brain had yet to develop this more complicated aspect of reward. Wanting, or "incentive salience," is, according to Berridge again, "the motivational incentive value of the same reward." In other words, wanting has nothing to do with sensory pleasure, nor does it necessarily have to be conscious. Rather it's the "motivational value" of a stimulus. Wanting is the silent math your brain calculates when you react to the juice, cocaine, or *Bizarro* cartoons.

You can like without wanting. "Dopamine suppression leaves individuals nearly without motivation for any pleasant incentive at all: food, sex, drugs, etc.," writes Berridge, but these subjects still show unconscious facial expressions of pleasure when given sugar. Subjects won't seek out the stimuli, but when given it they will react according to their own pleasure profile.

You can also want without liking. "During lateral hypothalamic stimulation, rats' facial expressions to a sweet taste actually became more aversive, if anything, as though the taste became bitter, although the same electrode made them eat." There is another study involving a bird who needs to nest but is not given any nesting materials. The bird had such a strong drive, or want, to build the nest that it used its own feathers, which couldn't have been fun.

The evolutionary psychology of liking versus wanting separates the two in order to allow the organism more flexibility in determining what is most rewarding. Wanting could have evolved to create "a common neural currency of incentive salience shared by all rewards, which could compare and decide competing choices for food, sex, or other rewards." In other words, liking helps you figure out if a single stimulus is good or bad, but you need wanting to compare two different stimuli. At last: the science of "soup or salad?"

Or wanting could have evolved so you could go after things without necessarily knowing if they would be good or not. Unlike liking, which is rooted in physiological reaction, wanting allows for speculation. Furthermore, it doesn't even have to be rational or guided by experience. Maybe New York would be a good place for me to make my fortune, even though I'm not cut out for it. Or maybe I should immerse myself in neuroscience, despite the fact that I have no interest or training in science and am barely able to make a living writing about television.

Now that Bill and my fate seemed to be linked, the difference between liking and wanting seemed important. He had always been a pleasure seeker. Even if he hadn't become a prescription-drug addict, he loved the sweet, downing Pepsi—which in our private lexicon was called "spruce"—any time he was awake. Then he discovered Breyer's Fudge Ripple ice cream, which he could consume a quart at a time; then came his Entenmann's phase. My senior year in high school, the year he really started to fall apart, he had a different favorite every month. I could mark the months in cherry cheesecake, marshmallow-topped fudge cake, chocolate chip cookies (a half-box at a time), all eaten with noisy, lip-smacking relish.

His other pleasures were also consumed with equal enthusiasm, but when it came to talking about his appetites, it was always within the confines of liking versus wanting. The line Bill gave me about how the prescription drugs were only psychologically addictive wasn't merely an effort at self-defense; he believed it. The pleasures of the world were "spicy," but he didn't need them.

As much as I had distanced myself from Bill, I could lay claim to the same appetites. Before Owen was born, I had indulged in alcohol, tobacco, gourmet food, gambling, and video games. In fact, I structured my life entirely around pleasure; I could write my own calendar of obsessions, from gin and tonics to small-batch bourbons to Marlboro Reds to Camel Lights to barbecue ribs to éclairs to blackjack to poker to craps to Xbox. I would never, ever put Heinz on a filet mignon, but that was a matter of taste.

Personally I had mixed feelings about the difference between liking and wanting. I could understand the idea of incentive salience for something like attention, but not for desire, or at least that was what I wanted to believe. I still hoped that my insight about turning into my stepfather wasn't true. There was one experiment that I could try, which, if the results were what I hoped they would be, might let me shake off Bill and get back to my original intent.

The day after I passed out in my clothes at the 2003 Society for Neuroscience annual meeting, I met a graduate student in Antoine Bechara's lab by the name of Nasir Naqvi. Naqvi was rocking a sixties aerospace-scientist look: charcoal-gray polyester slacks, a white shirt with a windowpane pattern on it, and big-frame eyeglasses. I could tell I was going to enjoy the presentation even before we started. What former smoker could resist a poster presentation titled "Emotional Responses Surrounding Puffs from Nicotinized and Denicotinized Cigarettes"?

"Did you derive pleasure from being a smoker?" he had asked me. I tried talking about the ritual, the cocktail-cigarette combination, but Naqvi was getting at something more elemental.

"Basically, what I'm going to try to convince you of is that a large part of smoking is what it does to your body," he had said. "Not your brain body but your *soma* body. In our department we make a big deal about it. Everything about you is physical."

Naqvi's whole thing was how drugs, and especially cigarettes, had an effect on the body. Basing his work on Damasio's somatic-marker hypothesis, Naqvi was investigating reward as something that is located *in the body* as much as it was a matter of neuro-hormones and neurotransmitters. In other words, the part of the brain that extended into the body through the central nervous system got an effect out of nicotine well before nicotine got into the bloodstream. Naqvi agreed to let me sample his experiment, which he said would illuminate the "somatic-hedonic" model of reward. Given my parents' problems with drug addiction, this is what I wanted to hear. I wanted my vices to be about liking with my body, not needing with my brain, which seemed to put me on track for failure and decay and who knew what other parallels between father and son. If I could locate reward in my body, then maybe I'd be okay.

I visited Naqvi shortly before the 2004 election. When I arrived at the University of Iowa Hospital lobby I saw: two largish women tying rag ends to a Pooh fleece blanket; a man with a fresh tracheotomy tube in his throat; and a man at the reception desk trying to unload a carton of eyes. These were more reminders that science is not all fun and light and future-leaning, but rather a place of consequence, something to be carried with you.

Earlier that morning I had gotten up at dawn to make the drive to Iowa. It felt like Bill was in the passenger seat next to me. I was trying to enjoy one of my favorite sights as a former smoker: people smoking in their cars in the morning. I passed a number of cars where the driver's-side window was cracked, the driver canted noticeably toward the gap, the tip of the cigarette creating a bridge between outside and in. I tried to take pleasure in the desperation of these car cigarettes (surely this wasn't pleasure) and the shame (why so afraid to stink up the car?) and the need of it, the cigarette like a straw delivering vital air rather than poison gas. But this seemed like something Bill would "get off on," as he would say, so I ended up listening to a Top 40 station to try to drown out his presence.

Naqvi greeted me in the lobby. He wasn't as fashionable this time, but then I learned that he and his wife were about to have a baby. He had other priorities now. Naqvi brought me to his office, where he gathered up his gear—a galvanic skin response machine, a stack of ashtrays, and packs of Quest cigarettes.

Knowing that I was a former smoker, Naqvi didn't feel right about letting me smoke for this experiment.

"You do still have the occasional cigarette, don't you?" he said.

I made a noncommittal gesture.

"You're not going to relapse on me, are you?" he said.

"You wouldn't want that on your head," I said.

"I wouldn't."

I didn't think I would, either, even though I had been a smoker for much of my life. I started smoking when I was nineteen. I was working as a telemarketer at Lou Harris and Associates, conducting telephone opinion surveys. There was a small, smoky break room in their office in Midtown, and because it would take too long to leave the building, everyone was forced to endure fifteen minutes of thick smoke twice per shift.

Bill and my mom smoked when I was very young, but they had quit a long time ago, and having grown up completely indoctrinated by the public school system into believing that smoking wasn't cool, I had always laughed at teen smokers. But the smokers at Lou Harris were different—not the kind of literate sophisticates Bill would have approved of, but fun nevertheless, and so I got my start smoking.

I smoked hard for the next fourteen years, then quit while we were trying to get pregnant. I had then been smoke-free for three or four years and didn't think I would ever go back. This experiment was going to be illuminating, but ultimately harmless. I had felt no wanting, and didn't even think there was any risk of liking.

Naqvi led me downstairs to the Environmental Exposure Chamber, where we were greeted by a respiratory therapist named Janet. There was a clean, cold-looking stainless-steel room off to the right that looked like a walk-in freezer. In the cramped office sat Janet, surrounded by tubes and hoses and tanks. The place looked like a scuba supply shop.

"How's the not smoking going, Janet?" said Naqvi.

"I bought a pack today," said Janet. She was medium-embarrassed.

The two of them then shared the dialogue of quitting, that conversation people who are trying to quit have when they think they're committed but aren't really. Janet missed the mini-packs that only contained ten cigarettes. Naqvi said he was uncomfortable with the dichotomy between smoker and nonsmoker.

"I'm not sure I can say I'm a person who will never smoke again," he said.

Smokers, I thought. *Could they be any more full of shit?*

We entered the Environmental Exposure Chamber. "This has nothing to do with my study," said Naqvi. "It just happens to be the only room in the hospital where you can smoke. The other place you can smoke in this hospital is a single room in the Psych Ward. But there are complications—the electrodes of the machine, the computer. You can see how that would feed into certain paradigms."

This room wasn't exactly without its tinges of paranoia. It was stainless steel, cold, humming with ventilation, with a slight echo that made me feel like I was in an Arctic research center. The room was normally used to test people who had been exposed to environmental hazards like diesel fumes or grain silo dust.

First there was, as there always is, paperwork. Normally Naqvi did his study on active, rather than lapsed, smokers, so he gave me a Fagerström Test for Nicotine Dependence, a Withdrawal Symptoms Scale, and Stephen Tiffany's Brief Questionnaire of Smoking Urges. The last form contained the immortal line: "Nothing would be better than smoking a cigarette right now."

"What you can do is smoke a cigarette right now," Naqvi said.

"This is kind of a training cigarette," I joked.

Naqvi handed me the stimulus, a Quest cigarette that he gets off the Internet. At 12:05 p.m., I had my first cigarette in four years. My inhale style was the same as before—a deep draw, hold, then exhale. The cigarette felt very natural. And rather than experiencing a racking cough, I took to the smoke quite nicely. I got a pleasant headache and not a shred of guilt. No experiment shame here. This cigarette was, in a word, delicious. This was definitely *liking*, but I also knew by now that I had to wait to see what the data said. Because the room was so cool and well-ventilated, everything about this cigarette was crisper and cleaner. I thought, *I'm smoking winter.*

Naqvi attended to the galvanic skin response (GSR) machine and explained what it would tell me.

"Instead of recording the emotional significance of what you say, it's recording the emotional significance of what you inhale," he said about the GSR machine. "Now, if you relapse because of this cigarette, I'm going to kill you," he said.

"That's fair," I said. "That's fair. I have no plans to do that. This is just for science. Purely for science."

"I've done that myself," he said, laughing. "I say to myself, 'I have to experience smoking again because I've forgotten what it feels like.'"

Then Naqvi gave me my instructions. He would light a cigarette and hand it to me. I was to hold it and wait for his cue, then take one drag. Then I was to rate the drags, on a scale of one to five, for pleasantness, strength, and desirability.

"Let's go," I said. "Let's smoke."

If you have never smoked in a climate-controlled, scientific fashion, I cannot recommend it enough. If you quit, it's time to come home. If you've never smoked, start. For all the fears of science taking the life out of something, here was an example of science putting life back in. I realized that so many of the cigarettes I had smoked in my life were cluttered. By coffee and alcohol and talking too much. But this was clean. I never realized smoking had so many *facets*.

The exceptions were the ones that were either unlit or didn't have any nicotine in them. The first test cigarette was a denicotinized model. I was immediately disappointed. I could instantly see what Naqvi was talking about, how nicotine affected the viscera. Even holding the denicotinized one, I could tell it was a sham cigarette. Bathed in its neutered smoke, I held it as if Naqvi had given me a disappointingly small serving of pie. Then I took a drag and felt nothing. Exhaling, I rated pleasantness, strength, and desire for more. All zeros. There was no heat in these heaters.

When Naqvi handed me a Quest with nicotine, on the other hand, I felt like I was smoking the somatic-marker hypothesis. The smoke hooked my nose and scraped my throat during the inhale and I was home again. It even felt different in my hand, and the curls of smoke reaching my nose made me realize this was one I was going to enjoy. I puffed (hard). I rated pleasantness (five). I rated strength (five). I rated desire for more (five).

We went through a few rounds of these, some with nicotine, some without. Even some that were unlit, just as a further control. Those were the most absurd and reminded me of the smokeless cigarette the tobacco companies tried years ago, and the ads featuring a guy toking on this piece of plastic while riding in a convertible. As if that were pleasantness.

"Hold," said Naqvi, but I was already on the smoke, puffing away.

"Sorry," I said. "I went right to the puff."

"A little eager there," said Naqvi. "We might have a problem."

We did have a problem. On the way home, I felt the wanting. It started even before I had gotten out of the hospital parking lot. It wasn't a nic fit so much as a reupping of what I had always found salient. When I saw people smoking in their cars, I was not as detached as I had been on the drive down. Rather than laughing at them, I watched their inhales, I followed the smoke out of the cracked window, and envied the blue air inside the car.

In Coralville, Iowa, I pulled over at a Chili's. Oxytocin is also released during eating—one researcher told me this was why people bond over food—so I loaded up on as much as I could hold. Since I couldn't have sex or smell my baby, I ordered up a cheap steak and some Southwestern Egg Rolls. When I left I was full enough not to want to smoke. It didn't hurt that I also felt bloated and a little sick.

Later, I got an e-mail from Naqvi with my results. They were a bust. "You'll notice that the visceral response does not differentiate between nicotinized and denicotinized puffs, while the strength, pleasantness, and desirability do," he wrote. "This is the normal pattern, and seems to be because visceral response does not track differences in hedonic impact that are due to the strength of sensations in your airway. However, I do have results from the group of subjects as a whole showing that the more someone craves cigarettes when they walk in, the more they react viscerally and hedonically to the bodily effects of smoking. This has implications for what cravings are 'about': the bodily effects of smoking."

I was confused, and a little disappointed. What Nasir seemed to be saying was that the data wasn't bearing out his hypothesis, but this couldn't be true. I *felt* the difference between the nicotinized and the denicotinized cigarettes, so why didn't the machine pick up on that? I looked over the graphs Nasir sent me. I couldn't read them, but I did realize something important: I hate data. There was one thread running through all of these experiments, even the ones I concocted myself, and that was that the data didn't match the experience. With the exception of fear conditioning, which helped illuminate the subtleties of fear, data was always telling me what I didn't want to hear. Now Naqvi seemed to be saying that in order for me to prove that my brain liked more than it wanted, I would need a full relapse. Unfortunately, my body was more than willing to oblige.

I didn't smoke for a while, but then I had a Camel Light the night Bush won the 2004 presidential election. I didn't much care for it. Out in the backyard of a friend's house, this cigarette tasted metallic and sour and left me feeling a little sick to my stomach. But the absence of liking didn't stop me from having another. Nor did it stop me from starting to secret-smoke when I would go out drinking. Then Liz's company got sold and we both started drinking wine and smoking cigarettes every night.

The cravings returned. For a while I fooled myself into thinking my smoking had experimental value, but then I realized I wasn't examining the facets of smoking, or exploring the intersection between my viscera and my brain. I was smoking—regular, workaday, completely unscientific smoking.

I wished Naqvi well on his quest to rewrite our perception of reward, but to me it didn't matter whether my desire to smoke originated in my soma body or in my brain. I was addicted again and all my rationalizations made me no different from any other smoker. If anything, the elaborateness and the intellectual reaching made me more like the very person I set out not to be. I had hoped that if my body merely liked, as opposed to wanted, then I wouldn't be like Bill. Now it was clear that my brain had wants, just like dear old Dad.

WHAT BRAINS NEED

I spent the next several months trying to figure out the neuroscience of Bill. I went back to the beginning, back to Seattle, then on to our dismal year in New Jersey, and finally to our family's disintegration in Forest Hills, and wrote down what I thought I knew. Then I took every memory and attitude and explanation to the University of Minnesota science library and tried to put it all back together, scientifically. Using their research database, I pored through papers about the effects of social stress on tree shrews, the inability of patients with orbitofrontal brain damage to experience social embarrassment, and how the genes that code for depression might serve a larger evolutionary purpose. Combining search terms like "amygdala" and "food addiction," or "mania" and "cortisol," I was looking for any connections that might shed light—I was Googling madness.

I wrote down important moments in Bill's decline as well as key neuroscience concepts on plain, five-by-seven note cards and taped them up on the bottom of the Murphy bed in my office. Some nights I would pull down the bed and the cards would hang underneath while I slept. When Liz and I were in the room together, we pretended the cards weren't there.

In some ways, atomizing Bill's descent into madness was almost getting me back to my original question about my brain. If I had undertaken this task ten years ago, what I was doing would have been psychological, but now these efforts were as much focused on the body as on the mind. If my stepfather lived somewhere in my brain, then I needed to know how deeply he had insinuated himself.

I quickly saw that I wasn't done with reward, even if my own problems with drugs, alcohol, and tobacco seemed small. The bar is high for addiction stories these days; if you don't wake up on a plane with teeth stuck to your face (or at least claim to), you don't have much of a hold on rock bottom. But the fact that I wasn't crawling, slobbering through a Tijuana gutter, didn't mean that the underlying mechanics of addiction weren't significant. My own addictions could be quiet and small and still tell me a lot about Bill and myself.

I had grown up understanding addiction in terms of what a scientist might call the "opponent-process" theory of addiction. This theory of addiction takes a homeostatic view. When people expose themselves to a strong hedonic stimulus over a long period of time, then an equally strong opposite response is generated in the brain. According to this theory, drugs of abuse hijack the neuroendocrine system. The brain adapts to the drug so that the drugged state becomes mistaken for the neutral, resting state. The brain then mistakes the absence of the drug as a threat to its new, maladjusted homeostasis. To the addicted brain, sober becomes a form of stress. Being straight is like being under attack.

Under this model it makes sense that if you can take away that stress, then you can ameliorate the addiction. This was essentially how my parents presented their enrollment in a methadone program. By design, the methadone was addressing their opponent-process problems, taking away the pain of withdrawal without giving them the pleasure of being high.

What I didn't know was that they were abusing their medication. Because they were professionals, they were trusted by their program, which let them collect their bottles twice a week rather than having to go in for a daily dose. This allowed them to hoard during the week, saving up vials that they would then double or triple up on come Friday night. It was a low-grade high, but a high nonetheless, and through some strange power of the methadone it allowed them to stay up all night, even if by two or three Bill had to slap himself to keep the fun going.

Come Sunday they were all out of medicine, and Bill was insufferable, often retreating to their bedroom, complaining of a migraine. He would bring a book and a saucepan to vomit into and draw the curtains and demand absolute silence. If the television or a conversation were one decibel too loud he would storm into the living room in a rage, asking us if we could even *comprehend* the pressures he was under at work or the demands of being someone of his intellect in a city of ignoramuses.

Then they would get a new batch on Monday and the beginning of the week would be fine, followed by increasing tension midweek as they started their savings program. Thursday was the worst day of the week. I thought this was the natural cycle of the working adult: Monday dread; the reluctant acceptance of Tuesday and Wednesday; the psychic pain of a tired Thursday, whose only relief came in Friday's promise of the weekend. But prescription drug abuse, including methadone abuse, is an invisible addiction. Because I never saw a drink, joint, or syringe in their hands, because the drug never made them anything but a little dozy, and because they were inherently unconventional people, I never imagined that their problems were far from over.

Their abuse of methadone also points to the incompleteness of the opponent-process theory. Just as reward is incomplete with only liking, addiction is incomplete without incentive sensitization. According to Berridge, the incentive-sensitization theory of addiction "does not deny that drug pleasure, withdrawal, or habits are all reasons people sometimes take drugs, but suggests that something else, sensitized wanting, may better explain compulsive long-lasting addiction and relapse."

The neurotransmitter that modulates this sensitization is dopamine. Once thought to be responsible for the pleasure derived from drugs of abuse, dopamine now is considered to mediate not only incentives and rewards, but what Dr. Nora Volkow, head of the National Institute on Drug Abuse and a leading addiction researcher, calls "general saliency." General saliency can be anything—aversive, pleasurable, novel, or simply unexpected—and this shift from pleasure to salience is significant. Now addiction, and by extension reward, means the brain isn't necessarily concerning itself with feeling good, but with being interested.

Under normal circumstances, the brain finds salience in natural reinforcers, which can be anything from sex to food to the pleasant company of friends. The dopamine D2 receptor seems to be at the heart of general saliency, along with its gene, the DRD2, which controls the number of dopamine D2 receptors a brain has. If the brain has a sufficient number of D2 receptors, then it finds natural stimuli reinforcing enough, but mice that have their DRD2 gene "knocked out," or removed, from their genetic sequence find natural reinforcers less interesting. The same effects are achieved with mice that have their dopamine D2 receptors blocked. For these mice, activities like running in a wheel don't provide the kicks they do for control animals.

Under the influence of certain drugs, however—for example, cocaine dopamine levels in the brain dramatically increase to levels five to ten times higher than the levels caused by natural reinforcers. Under the incentive-sensitization model of addiction, drugs of abuse replace natural stimuli. Volkow posits that drugs of abuse raise dopamine levels so high that natural reinforcers can "mimic but not surpass in intensity and duration the DA [dopamine] increases triggered by phasic cell firing." For the addicted mouse, the drugs are much more like it.

In humans, studies have shown that drug addicts have fewer D2 receptors than controls. "What happens if someone has fewer dopamine D2 receptors?" Volkow said in a public lecture at a Society for Neuroscience (SFN) symposium I attended in 2003. "The dopamine is immediately recycled. Fewer receptors means less interaction. The person is then less sensitive to natural reinforcers. The addicted person learns that natural reinforcers are no longer salient and also learns that drugs of abuse are."

Maybe this was what was happening on Bill's hangover days, which were less a matter of the pain of withdrawal than of the lack of salience of his nondrugged state. This seemed to fit with the overall pattern of his life, which always looked like an exercise in hedonism, but was in reality a search for salience. He needed drugs and huge mouthfuls of food and zany behavior because the straight world wasn't enough. Perhaps on account of his genetic makeup the straight world could *never* be enough. There is even a name for having too few dopamine D2 receptors: reward deficiency syndrome.

So, if there was such a thing as reward deficiency syndrome, then were the people who weren't vulnerable to drug abuse benefiting from some kind of reward *sufficiency* syndrome? In the same SFN talk, Volkow spoke of an experiment where she gave a mild stimulant (Ritalin) intravenously to healthy nonaddicts as well as to drug addicts. Using radioactive markers, she also counted the number of dopamine D2 receptors in both subjects and controls. Fifty percent of the overall subjects reported liking the experience and 50 percent found it unpleasant. The ones who did like the drug had "significantly lower levels of dopamine D2 receptors than the subjects who described it as unpleasant." According to Volkow, there seemed to be "an optimal range for DA D2 receptor stimulation to be received as reinforcing; too little may not be sufficient but too much may be aversive."

Volkow then said that "drugs aren't just about drugs. It has to do with the individual biochemistry in the individual brain." To me this sounded like this: "Drugs feel good to people who like taking drugs." But now this seemed significant. Drugs feel good to people who like to take drugs, but also, drugs don't feel as good to people who don't like to take drugs. Not only were addicts genetically more vulnerable, but nonaddicts were genetically less vulnerable. What the drugfree saw as high moral fiber was maybe a matter of dopamine D2 receptor wealth—chemistry, not character.

Drugs of abuse can also change the brain's dopamine receptor makeup. Chronic alcohol abuse will actively lower the number of dopamine D2 receptors in the brain. Just like people, neurons can be lazy. Because alcohol increases dopamine levels in the brain, the brain sends the signal to shut down what it sees as extra receptors. If someone else is doing all the shopping, then the brain sees no need to go to the store.

As Bill got closer to his breakdown, he lost more and more interest in the basic things a person needs to do to take care of himself. I was now the only one who did the grocery shopping and the family laundry, because if I didn't do these chores they wouldn't get done. I would try letting the house go, just to see how bad they would let it get, but I always blinked first. They didn't care about eating or cleaning or returning movies or showering on the weekends. They pulled themselves together every day to go to work, but only barely. I would talk to them about how they needed to get their act together and they would nod as if I were wise. "You're right, lad," Bill would say, but while his mind agreed, his brain didn't see the point.

Of course, I didn't have access to Bill's brain. And even if I did, knowing his genetic makeup might not have provided the answers I wanted. The more I searched for a genetic answer, the more elusive the answer became.

Genes are like a movie script. There are some basic plot elements that remain firm, some scenes that you simply must have. But the way that script is produced can yield a variety of results. You can give the same script to different directors and casts and it will turn out differently each time. You can have a great cast and a terrific premise but a hack director and a focus-group ending that results in a steaming pile. Or you can have a relatively conventional story, and then that washed-up nobody, that forgotten television actor, gives the performance of a lifetime and everyone says, "Where did that come from?"

Having opened up Bill's reward network to genetic scrutiny, I felt obligated to do the same for myself. But how? First there was the problem of my genetic inheritance. I was in denial about my biological father, who came back from Vietnam when I was one or two years old, suffering from what I had always thought was schizophrenia but was later told by my aunt was post-traumatic stress. (I had been told by many doctors, psychiatrists, and counselors that since schizophrenia hadn't manifested itself yet it was unlikely that I was ever going to lose my mind in that particular way. As a result, I had put that worry behind me years ago.)

But if I was going to look at the reward genes in terms of my biological inheritance, was that going to be enough? Even if I could get a peek at my biological father's and my mother's genes, I would then have to factor in how my own upbringing affected how those genes were expressed, plus whatever behavioral patterns I had created for myself in the twenty years since I had left home. All this left me feeling like someone's sick idea of the perfect nature/nurture experiment.

I tried to embrace the genetics of my whole brain. My dream was to cook up an entire genetic array, a full battery of tests that would give me a complete genetic profile—with an emphasis on attention, anxiety, and reward—but the people at the National Institute of Mental Health were lukewarm to this idea. My dealings with the public relations department there were going very well until I suggested that they give me a broad genetic profile. I was quickly shut down. The communications officer said that what I was proposing was "still the stuff of quacks and science fiction writers." The quack part I could handle, but the bit about being a science fiction writer? That smarted.

I was also relieved. Now that I had Owen in my life, the idea of injecting a radioactive tracer into my body to determine my dopamine D2 receptor count seemed risky. The potential benefits of knowing if I had reward deficiency syndrome didn't outweigh the risks of undergoing the scan. Even if it was deemed safe, it still sounded pretty scary. This left me to speculate, which I realized I had been doing for most of my research, anyway. A little more couldn't hurt. In the end, I decided to focus on my drinking. Apart from cigarettes, this was where I had the most reward experience, and when I read about chronic alcohol abuse, the effects seemed to correlate nicely with my list of complaints about my brain.

Alcohol does nasty things to the brain, both in terms of its overall physiology as well as in terms of its brain chemistry. Chronic alcohol abuse leads to overall brain shrinkage, atrophy of nerve cells, and a decrease in brain metabolism, especially to the cortex. At the most extreme end of the spectrum, this results in Korsakoff's syndrome (KS), a memory disorder where the sufferer has problems creating new memories. (Marlene Oscar-Berman notes in her paper "Impairments of Brain and Behavior: The Neurological Effects of Alcohol" that "someone who developed KS in the 1960s might believe that the [current] President of the United States is Dwight Eisenhower or John Kennedy.") But new evidence from autopsies and brain scans showed that non-KS alcohol abusers still show damage. "The once commonly held view that alcoholics without KS are cognitively intact has been abandoned," writes Oscar-Berman in the same paper, "in light of accumulating evidence that cognitive impairments (and associated changes in brain structure) can occur in alcoholics who do not exhibit obvious clinical signs of anterograde amnesia [i.e., the amnesia of recent occurrences]."

I didn't think I was an alcoholic, but then again there is no scientific definition of alcoholism or alcohol abuse. According to the National Institute on Alcohol Abuse and Alcoholism, moderate drinking is no more than two drinks a day for a man and one drink a day for a woman, with no more than four in any single day and no more than fourteen a week for men, and no more than three a day and no more than seven a week for women. Anything beyond that is considered abuse. As for alcoholism, an NIAAA spokesperson told me alcoholism is a pattern of harmful drinking where the drinker recognizes the harmful effects and yet persists. "We don't yet understand it as a disease," she said. "We know it by its symptoms."

I could no longer ignore alcohol. Early on in 2005, I added another investigatory thread to my efforts to understand the neuroscience of Bill—my own booze history.

I started calculating my alcohol intake. Then, in the interest of being complete, I quickly opened my inquiry to include marijuana, cocaine, LSD, hallucinogenic mushrooms, opium, airplane glue, nitrous oxide ("whippets"), and codeine (in cough syrup and Mexican pill form). The idea was to make a cold, objective inventory of every substance I had used, and then take this information to the U of MN science library to see if the databases there could help me tell the story of what fun had done to my brain.

First up: marijuana. I started smoking pot the end of my junior year of high school. By fall term of my senior year I was smoking almost every night. After my parents would go to bed I would smoke a bowl, go to Roy Rogers to buy fried chicken, then come home and watch *Late Night with David Letterman*. Then, in my freshman year of college, I switched to bong hits, which were administered multiple times a day the entire year. My sophomore year I cut down considerably, then switched to very occasional use. I don't know how they measure marijuana intake. In tokes? In bowls? In cubic feet of bong smoke? For my purposes I decided to use the hit as the unit of measure. My calculations said 400 hits for first year as pot smoker, then 3,500 hits my second year, then maybe 1,000 hits the third year. Since then, maybe a total of 100 after that.

Total hits: 5,000.

When I looked at the science of marijuana, I found good news, or at least the kind of good news that comes when there is a scarcity of research. The abstract to Robert Block's 2000 paper "Effects of Frequent Marijuana Use on Brain Tissue, Volume, and Composition" offered this bit of cheer: "Marijuana users showed no evidence of cerebral atrophy or global or regional changes in tissue volume." Block had earlier noted in a *JAMA* editorial that marijuana hadn't been studied enough for there to be anything conclusive. I took inconclusive to be a rubber stamp that said: Pot okay.

For many of the drugs (such as opium, LSD, and mushrooms), the problem was a complete lack of information. Because research is funded almost entirely on the basis of its relevance to addiction, the small-potatoes drugs don't get much attention. There were hundreds of papers on alcohol and cocaine, but an underdog like mushrooms wasn't deemed enough of a problem to warrant any attention. This was fine. My total cocaine, opium, LSD, and mushroom intake didn't amount to more than a couple dozen encounters. Conclusion: Dabbling okay.

This was all a way of putting off looking at alcohol. I was fourteen when I had my first booze-up. My friends and I drank brandy from a bottle secreted in a black dress sock. We ran around the neighborhood. We talked about girls. I humped a tree. In college the drinking began in earnest. I calculated an average of 24–30 drinks a week, or 288–360 drinks per trimester, or 864–1,080 drinks per school year. During the summer I was probably drinking slightly less than that, which put me at 6,000 drinks to finish college. In my twenties I still probably drank about 1,000 drinks a year until I moved in with Liz, who was appalled at the fact that I went out almost every night. We moved in together when I was twenty-eight or twenty-nine, which was after another 6,000 drinks, for a grand total of 12,000 drinks to get me to the point where I was living with a woman. Then a pretty big drop-off in consumption. Another 2,500 our first five years together, then 750 in the past five years. With Owen in our house there was almost no drinking at all, especially in those first months when even the thought of drinking filled me with fear.

Total to date: 15,000 drinks.

Is that a lot? As a NIAAA report noted, the link between alcohol and brain damage is unclear, but one study did show that social drinkers who consume more than twenty-one drinks per week could suffer from cognitive impairments. "Long-term, light-to-moderate social drinkers have been found to fall into this category as well," the report said, "showing cognitive deficits equivalent to those found in detoxified alcoholics."

I hopefully hadn't consumed enough to cause large-scale organic damage, but regardless, the brain seemed so much more delicate to me now. I started to wonder if my lifetime of drinking didn't explain some of my attention problems or working-memory problems or even my stress. You don't have to be an addict and you don't have to have dramatic impairments to be damaged. You're a little slower, a little softer, only a few tenths of a percentage point off, but it's enough to make a difference. If anything, it's these little losses that hurt the most because you don't see them coming until it's too late.

Also, Bill was starting to look better and better. He made a good choice sticking to prescription drugs, at least cognitively speaking. His addicted brain had its reward deficiencies and sought salience in drugs, but he had avoided the organic alcohol damage. He stayed sharp. When he chose to concentrate, he could muster incredible attention to detail, as well as the ability to put information together and make associations. And, at least back then, he remembered everything.

Killing brain cells. That was always the joke in college. Now I understood that killing them would be mercy. Instead the brain lives on, but it gets retrained. Imagine if you had to unlearn English. Not stop speaking it, but actually unlearn it, erasing it from your mind, from your brain, so you were incapable of speaking English and instead had to pick up French or get really good at pointing. This was how drugs of abuse seemed to work. Even if there was no organic damage. It created an entirely different language for the brain.

In my twenties I lived under the illusion that the methadone program, which my parents eventually quit, had worked. Even after they left New York and moved to San Francisco and I started getting frantic calls asking me to wire them money, or when I would call them for four straight days and the phone was never picked up, or when I would go to visit and their apartment was piling up with clutter, or when Bill finally lost the last job he would ever have—throughout all that, I never thought it could be drugs. I now knew that as much as it was possible to beat addiction, the neuroscience of drug dependence made it unlikely for some people. Some people just probably weren't ever going to get better.

As for myself, I used to think in terms of addicted and not addicted, of gluttony and moderation, but the learning-brain model revealed a gray scale that was more nuanced (and, in my mind, more devious). At least with addiction you could point to a single villain, but I had unwittingly spent the past twenty years of my life training my brain to respond to certain stimuli. A glass of wine or a weekend in Vegas might not damage me, but it was subtly replacing other options. The time I was spending drinking, smoking, gambling, and watching Comedy Central was taking time away from other pursuits, like reading or learning a musical instrument or whatever else is supposed to be good for you. I went back over the Brain Logs, my diary of all the bad TV and fast food, and cringed. I thought I was watching *Terminator 3: Rise of the Machines* with ironic detachment, but in reality the crap I was feeding my head meant something. The brain was *always on*. There was no work time and leisure time. No wonder we were so messed up as adults. You spend the first half of your life unwittingly becoming who you are and then spend the rest of it trying to untangle the knot that is you.

There is a reason why scientists don't study themselves—it's too painful. When applied to the self, the brain is just plain depressing. In all my research I only encountered one piece (one piece!) of unequivocally good news: lactating mice would rather breast-feed their young than take cocaine.

There wasn't a single system or function or protein that couldn't be used for ill as well as for good. Often the news was just awful. One report said that half the population will be declared mentally ill at some point in their lives. Elsewhere I read that, every decade, my brain undergoes a natural 5–6 percent decline in dopamine D2 receptors. If you calculate that out, starting for the sake of argument with one hundred receptors at age twenty, then by age seventy you have seventy-seven. My life in retirement was going to be a quarter less salient than when I was young.

I was also watching Owen with increasing anxiety. When I first started researching the brain I promised myself that I wouldn't train the Eye of Science on him. I didn't want to dehumanize our relationship. But now I couldn't help myself. By the spring of 2005, Owen was becoming more and more of a person. He was walking and starting to talk, and what had started out as likes were turning into *needs*. He was beginning his own long, slow process of training up his reward network.

Now that I thought I understood reward, I felt it was my job to head off any future disasters. I came up with some guiding principles for his upbringing. It boiled down to three rules: no sound chips in his toys, no licensed products (not even Sponge-Bob), and only one hour of television a week. Once in a while a sound-chip toy would sneak into our house and I was terrified at its effects. Normally a content, creative child, Owen would turn into a button junkie, pressing a toy to make it roar or sing over and over and over again. If I tried to steer him away from the toy, he would get agitated. He liked books, but the more he learned how to talk, the more he expressed a strong drive to watch television. Sitting on the couch, two fingers in his mouth and the thumb from his free hand stroking his upper lip, he shouted out demands for *Teletubbies, Bob the Builder, Miffy*, and "just a little" *Monsters, Inc.*

I would try to tell him that we only watch TV on weekends, or sick days, or when Mommy and Daddy are very, very tired, but he only matched my reason with shouting.

"I! Watch! Little! Monsters! Inc.! Daddy!"

I felt sorry for him. How could I explain that if he watched as much television as I did growing up, no matter how smart he was, like his dad he would always have a shit-for-brains streak to his thinking. He would always wonder what might have been if he had learned to be more rigorous with his mind. How could I tell him that if he didn't try to own his brain, it would end up owning him?

NO PRESCRIPTION DRUGS

The summer of 2005 marked a new low with myself and the brain. Between the brain, Bill, and the baby, I was suffering from a kind of premature midlife crisis. Neuroscience had me feeling that, neurologically speaking, my life was essentially over. I had nothing more to look forward to than decreased salience, increased stress, and the continued accumulation of unconscious fears. I wasn't about to buy a motorcycle or run out for a pack of cigarettes and never return, but I could see the appeal.

Neuroscience had also taught me that big, sudden moves were futile—the brain changes incrementally over time. Scans of professional musicians show physiological changes to the music-processing parts of their brains, but only after years of practice and performance. If I wanted to change my reward network or hone my attentional skills, then I would need to give myself twenty years.

Or I could take a pill. I had been in and out of therapy for the past ten years, but I'd always stopped short of any kind of pharmacological solution. I was always afraid that an antidepressant or an antianxiety drug might change the essential me. Now I felt that whatever essential me there might be was constantly shifting anyway. For the first time in my life I felt like a pill might actually make my brain feel like itself again.

This was Bill's hope when he first went to Dr. M. In 1985, when Bill told me that he had been diagnosed with bipolar disorder, he presented it as purely a medical issue. Like the drug problem, it wasn't couched in terms of its impact on our family, or in terms of his overall well-being—his mental illness was something that was only happening to him. Bill's commonsense dualism told him that the vehicle his self occupied needed repair. Fix the vehicle and its driver would be fine.

Dr. M. first prescribed lithium in an effort to control Bill's mood swings. After the lithium, there was a host of others, with names that sounded so strange and made-up that I couldn't remember them even when Bill repeated them again and again. The only thing that seemed consistent was the side effects. The drugs gave him insomnia. The drugs made him fat. The drugs made him sluggish. There was always a point when Bill thought the drug or the dosage was right, but they never seemed to make him better.

He also made taking psychoactive drugs seem fun. Bill liked to read the list of side effects and riff on them. When Dr. M. put him on Parnate, Bill made a list of cheeses he couldn't eat. He kept the list in his wallet, and would use it both for reference and for laughs.

After Bill's psychotic break I started to see Dr. M. myself. Even though I had seen the effects antidepressants had on Bill, I didn't question Dr. M.'s idea to put me on something. Those early drugs, the MOA inhibitors and the tricyclics, were blunt instruments, and I'm not even sure if I was clinically depressed, but Dr. M. seemed to think a little nortriptyline coupled with Ritalin would be the right cocktail for me.

By this time I was attending a small midwestern liberal arts college. That was Bill's idea. At one point he tried to convince me to go to West Point, because then it would have been free, but ultimately he clung to this fantasy of the liberal arts education. "You're going to learn how to learn, lad," he would say to me, before launching into a two-hour talk about the history of the liberal arts degree. There were times when I think he still hoped that one day I would study Latin.

I started taking the drugs shortly after Bill's psychotic break. I hated the way they made me feel. The nortriptylene gave me these epic dreams, with weather and color and multilayered sounds. I woke up once and walked into our kitchen in Queens. Cursing the dirty knife, covered in marshmallow and fudge crumbs, I then started shaking as the walls around me started to melt. When I wasn't experiencing fitful sleep or low-grade hallucinations, the drugs made me sluggish and cotton-mouthed. The Ritalin, however, was fun. Once back in college, I took six or seven at a time, often with alcohol, and dispensed them at parties to my friends, who called them "zippy pills." But I was still depressed. And paranoid—I thought people were talking about me in code. I stopped leaving my dorm room, except to bring food up from the cafeteria. By spring term I was off them and swore never again.

Still, during my research into reward, I got interested in a drug called Adderall, which was prescribed for both children and adults with ADHD. The great thing about Adderall was that it was time-released. Rather than taking Ritalin several times a day, you took one Adderall and that, presumably, would take care of your frazzled mind.

I went to see my general practitioner. My doctor used to have a kindly beard, but now he was clean-shaven, and while he appeared younger, the new look hadn't sapped his authority one bit. I told him about my book and how I was doing experiments on my brain and he nodded warmly. Then I told him about my continuing interest in attention, and that I was hoping he would prescribe me Adderall, "for journalistic purposes."

"Well, Dr. Jekyll," he said, in the kind of theatrical tone you would expect to hear at a mystery dinner.

I was conscious of not displaying what clinicians call "drug-seeking behavior," the sort of headache-exaggerating and pain-faking my parents would employ to get their drugs. So I downplayed the request as best I could.

"I only want a few," I said. "A handful. Less than a handful. Less than ten."

"And how do you plan to be objective?" he asked.

I had no answer for that, though secretly I knew I was pretty close to done with objectivity. If objectivity and I couldn't get along yet, I doubted we were going to start a torrid affair now. I could have thrown out some nonsense about planning to do two-letter cancellation tests, or the Digital Symbol Substitution Task, but instead I admitted that I was primarily interested in the subjective experience.

My doctor consulted my file and nodded. "A lot of people find that they reach a point in their life when they're not having the kind of success they want," he said. "One of the pharmaceutical sales reps who visits us has a slightly worse profile than yours and he finds Adderall very effective." I looked at my file, which from where I was sitting was upside down, and saw a printout of my old ADD screener. I had forgotten about that. Here I was two years later, and it was *data,* safely nestled in my thick file of annual checkups, along with the biopsy on my neck and the upper GI they gave me when I had a preulcerous condition. My doctor, who has taken care of me for more than ten years, wasn't thinking about my research. He wanted to help.

"I want the minimum," I said. "The *minimum*."

Before I knew it, my doctor had picked out what he deemed to be the appropriate dosage and prescribed sixty 10 mg tablets of Adderall. After that, a brief, friendly chat, and then I was off. I walked through the waiting room, which was filled with people who had certainly not come here for journalistic purposes. The word "score" popped into my head.

Later that day, I picked up my pills. The info sheet attached to the bag described Adderall as a "D-Amphetamine Salt Combo." It sounded delicious. Inside the bottle were oval-shaped blue pills that reminded me of the deep end of a swimming pool. The cost was $12.21 with our health insurance, marked down from almost $100. The warning label—which I read in great detail—said Adderall was merely "habit-forming." Before I met my brain, I would have shrugged this phrase off, but my knowledge of reward and addiction and the chemical depletion of dopamine receptors made this casual phrase seem freighted with danger. The label also said I was not supposed to take this medication if I was experiencing emotional instability or if I had a history of drug abuse; but then, that was the whole reason I was here.

Now it was a matter of designing my experiment. This time I wasn't even going to try to compete with real scientists. Researchers who conduct drug studies with ADHD medications employ elaborate methodologies. The Lab School Protocol, for example, involves kids attending a mock school, complete with classes, recess, and "other activities," to test the efficacy of drugs like Ritalin and Adderall. For their paper "Efficacy of Mixed Amphetamine Salt Compounds in Adults with Attention-Deficit/

Hyperactivity Disorder," the authors (Spencer et al) did a "7-week, randomized, double-blind, placebo-controlled, cross-over study of Adderall in 27 well-characterized adults satisfying full DSM-IV criteria for ADHD of childhood onset and persistent symptoms into adulthood." Research subjects were given a SCID with a Kiddie Schedule for Affective Disorders and Schizophrenia for School Age Children (Epidemiologic Version), as well as the Wechsler Intelligence Scale (Revised); the Wide Range Achievement Test (Revised); the Hollingshead Four-Factor Index of Social Status; the Clinical Global Impression Scale; the Hamilton Depression Scale; the Beck Depression Inventory; and the Hamilton Anxiety Scale, the Continuous Performance Test, the Stroop Test, and the Rey-Osterrieth Complex Figure Task.

My plan: take a pill and go to Arby's.

The Arby's in my neighborhood is a blight. I picked it as the testing ground for the first time I took Adderall because I couldn't imagine anything inherently salient about it. I was very neutral to it, neither hating nor loving it. It was a place in my neighborhood that was a blank slate to me. I reasoned that there was nothing at Arby's that I would want to attend to, and nothing I wouldn't want to attend to. This made it the perfect place to test out Adderall, which should have enhanced my ability to attend to meaningless stimuli.

I was going to focus on inhibition. The goal would be not to look at people. If Adderall worked, then it would help me avoid distractions, which at a place like Arby's could only mean the other patrons. I should be able to attend to the least salient material possible, the flecks of color on the carpet, the pattern of the ceiling, the exit sign, the food. My plan was to eat lunch (no caffeinated beverages) and keep it simple. The only thing I had to worry about was whether or not the Beef 'n' Cheddar Sandwich would confound the results.

First I went there clean, as a control. I stayed pure the night before, avoiding alcohol and tobacco and sugar. The morning of the experiment I went for a run. When I arrived at Arby's it was as bland as I had hoped. I ordered a Beef 'n' Cheddar Sandwich, a side of curly fries, a side of Jalapeño Bites with Bronco Berry Sauce, and a Sprite, then sat down and paid attention to where my attention went.

As I expected, even in a place that held no interest for me, it was hard not to attend to things. I tried to focus on the boomerang pattern on the table, a Keith Haring image hanging on a support beam, the sailboat art, and the photographs of the St. Paul and New York City skylines at dusk. When you really study your attention system, you realize just how much your brain is sliding over its environment. I felt like a nervous bird, even though there was nothing to be nervous about.

My prediction that it would be hard to inhibit the stimuli of other people proved correct. Social stimuli are obviously important even to a prehistoric brain. There was a shaved-head swell in a pink shirt, a girl making eyes at her boyfriend, an old couple drinking coffee. Even more distracting was all the human movement. People entered and exited the ordering area; they carried their trays through the dining room; they ambled out to their cars in the parking lot. I really tried to look at the fake woodgrain paneling on the wall next to my table, but then someone reached for their soda and my eyes were all over them, even though I had plenty left to drink.

Experiments often have what's called a mask. During one experiment that involved word lists, I had to do simple mazes in between trials, thus blasting my mind free of the previous batch of words. So, a few days later, under the same clean living conditions, I went to Einstein Bagels, a chain bagel store also in my neighborhood. It's not the best bagel, but obviously I gave up on the bagel as an art form in 1986, when I left New York. I did the same exercise. Bagel dog, high school kids on break from sailing camp, potato salad, local alternative weekly, old man with weird glasses, bagel dog. I counted eighteen attentional shifts in five minutes, with much of that time dedicated to note taking. My mind had been cleared.

Then, on July 1, 2005, at 11:47 a.m., I took one 10 mg tablet of Adderall. It was sweet on the tongue.

I was fine on the walk over, and then it hit me, in line, a tight, electric feeling that reminded me that this was after all a mild stimulant. In *The Doors of Perception* Aldous Huxley tumbled into the beauty of a flower on the desk of his office, and when I read this I thought: *Of course.* Of course when you're on peyote a flower will reveal the infinite. Would he have found the same with a can opener? Or a CD offering fifty free hours of America Online?

But now I was not so skeptical. All the extra dopamine was making everything extra salient. My vision felt sticky. It was like a photorealist painting in which all depths of field are preternaturally sharp. The Keith Haring print popped off the post like an optical illusion. I could focus on it and take in the sign for the Spin Cycle Laundromat across the street. I read the sports page like I'd never read it before in my life; a piece about the Minnesota Twins' need for a "stable infield" felt like it was written by my favorite author. I marveled at the corn-syrup shine of a jewel of Bronco Berry Sauce on the container lid. The boomerang patterns on the table, the heretofore boring carpet flecks and ceiling tiles, the grains in the wood paneling. I could look for as long or as little as I liked. The Arby's was still a hideous place, but all these details made it beautiful.

Inhibition of stimuli was no problem. At the table next to me, a trio of college students, two girls and one boy, gathered around their trays to listen to one of the girls remind them about "that time." Normally I would be all ears for this kind of story, the cataloguing and time-stamping of various alcoholic beverages, the proud depletion of metabolized glucose and the steady shrinkage of the PFC. With the Adderall, I could shut them out and focus on the most boring thing in the my field of vision. Of course, there was a price. The Beef 'n' Cheddar, which was bad before, was now awful; my sense of taste—and, even more, vision—rendered it completely unappealing. Thanks to the Adderall, I now understood, with great precision, what a miserable sandwich it was.

Then the drug kicked in a little more. By 12:30, I was starting to space out a little. I found myself staring at the whorls of an old man's Ham 'n' Cheddar. The way the meat folded, endlessly, like the universe—*like the human brain!*—I could have stayed there all day. Instead I went to my local record store, where I almost fell into a CD bin, caught in the shine of the wrapper for a Bright Eyes CD, when I wasn't interested in Bright Eyes. Then I also almost got hit by a car because my attention was so focused I wasn't being diffuse when I needed to be. Finally I had to buy a pair of sunglasses because I couldn't handle the brightness anymore.

When I first saw Liz, she wanted to know how I felt. I tried to explain in terms of coffee or nicotine but neither made sense. We sat out on our deck on a beautiful, perfect Minnesota summer day that felt extra sharp and clean and clear. I told her about reading the sports page and being amazed at my new abilities.

"So it lets you do things you don't want to do," she said.

"Exactly," I said.

She gestured toward a brown box that had just arrived from UPS and said, "Then how about hanging that light fixture?"

I did change that light fixture, and promptly, and I got a lot more done, too. I told myself it was for the sake of science, but ultimately I kept taking the Adderall because the "experiment" was such a success.

The biggest difference in my life was how productive I became. The second time I took Adderall, my vision was less sticky and I wasn't in danger of getting hit by a car or falling into a CD bin, but I was able to focus like I remember being able to focus in high school, back when my brain was relatively pure and uncluttered. Even after a night of drinking, I was able, with the help of Adderall, to process incredible amounts of information. I could read a half-dozen scientific research papers without getting up once. Difficult concepts such as Hebbian learning seemed like comic strips in their simplicity. Anything easier insulted my genius.

There were times when I still jumped from task to task, from article to chores, but I was less stressed about it. I didn't hate myself for being a little flighty. I reasoned that it was because the Adderall had beefed up my prefrontal cortex overall. I now had more cognitive control, which made me less emotional, less stressed, less anxious. My amygdala, once my ruler, was now no match for the higher functions. Even if the coolness of reason is an illusion, subjectively that was how I felt, and there were times when I would go to Malcolm Gladwell's website, just to enjoy looking at it, without feeling any jealousy at all. I could even feel a little sorry for him, imagining how hard he tried to make everything perfect.

Was I high? Yes and no. I was at high's doorstep and I had my nose pressed against the screen door. Addiction researchers have found that the speed with which a drug effect manifests itself is a crucial part of not only the high but also the severity of the addiction. Adderall is time-released, which meant there is no initial bursts, or at least not the kind associated with a recreational drug. Still, I felt I was subtly peaking about an hour after I took the pill. The skin on my skull tightened and for about five minutes it felt like someone was popping bubblewrap inside my head.

And I was high, but, again, not in the sense of pleasure. Bill used to talk about "exuberance." That was how he described his manic phases, and he was always disappointed in my mom and me for not relishing his exuberance. On Adderall I got a little taste of his mania. The feeling of being jazzed up all the time, especially in the morning. In fact, everything was more enjoyable, more salient. Music sounded better, video games delivered a more visceral punch—I felt like I was twenty-five again. Life was interesting.

There were times when I felt so sharp it almost felt unfair. Ethicists studying the societal impact of neuroscience, worry over the neurological haves and have-nots. I was most certainly a have. One day, I went to traffic court to resolve a fender bender. I popped a full 10 mg pill, let it kick in, and then dressed like I felt: flared jeans, a blue blazer, and a faded red T-shirt that said, "I'm a Pepper." Up in front of the judge, I was cool, clean, and effective, while my opponent dithered and mumbled and finally lost his temper. "I couldn't have hit him that hard!" he said. I looked at the judge, doing my best not to sear his retinas with my laser vision, stood by my story, and won.

Then my brain started to adapt. By the third or fourth week, the bad habits crept back in. The wandering, the video-game playing, the long lunches. I still got the benefits of increased productivity, but the feeling wasn't the same, and after a while Adderall stopped being an experiment and became my routine. I took either a half or a whole Adderall every morning when I first woke up, then a day of brain research, then time with Liz and Owen, then two to four cigarettes and a glass or two of wine, then a little more brain research, then bed.

I felt like I might be getting into a situation with the drug—nothing alarming, but I found that when I didn't take the Adderall, I missed the energy. I started to rationalize. I told myself that in order to get my old self back I needed to embrace a more chemical me. This was good for my family, right? The prehistoric brain was no longer enough. Evolution was too slow. Whatever success we had as a species has never been on account of our patience—if we want to build a log cabin, we don't wait for the trees we need to fall down on their own.

I also convinced myself that there was something right about this way of living. I have always felt physically better when I'm exercising, eating right, and getting plenty of sleep; when there is less sugar in my diet and when I'm reading good books and communicating with Liz and being loving toward Owen and taking time out for myself to go for walks and meditate and do yoga. But after a few cigarettes, or a drink or three, or half a pie, I have always felt more like myself.

By the second month, I still felt good on Adderall, but I was also a little off. I would stop by Owen's day care and just talk people's ears off. I found myself talking and not being able to stop; the part of me that watches the rest of me felt out of control. There were also moments while spending time with Liz and Owen when I felt a little distant. It wasn't that I was cold, but the extra cognitive control did seem to put a thin membrane between us. The natural reinforcers still had their effect, but they were muted. And then there was the fact that apart from the first time I took Adderall, Liz didn't know about any of this. Yet, somehow, I was always able to rationalize these negatives away.

This is how prescription drug abuse happens: smart, educated people self-diagnose, then meet with doctors who want to help. I had always thought of Bill's drug use in terms of his arrogance and self-blindness, but now that I was doing it to myself, I could see how he and my mother got hooked in. Like me, they had a prescription, from a doctor, which bestowed legitimacy. Like me, they felt the effects largely internally. A little more energy for me, a little less pain for them—as long as other people didn't notice the difference, what could be the harm? Did I not have a right to manage my own brain? Didn't we all?

It was thoughts like these, internal monologues that happened at the supermarket or while walking around the lake with Owen and Liz, that made me realize I needed to stop taking the drug. If these were such good arguments, then why did I have to keep them secret? But first I was going to use the drug for one last bit of good. I had looked at the history of Bill falling apart in New York almost in its entirety, but I had shied away from the night of his psychotic break. Since my journey through the brain had destroyed so many of my assumptions about Bill and myself, it was only fair to reprocess everything. It was time to go back to Amsterdam Avenue.

THE NEUROBIOLOGY
OF CRAZY

I woke up at 3:00 a.m. to the sound of *Star Trek*. My mom and my half brother were sleeping in my parents' room nearby, so I tiptoed into the living room, an absurd gesture since the volume on the television was loud enough to be heard in the tennis stadium across the street.

I found Bill on the couch, wearing shorts and an old wife-beater. Thank God he wasn't wearing his towel. In the final days of the decline, Bill took to wearing a small, thin, powder-blue towel that fastened around his waist with a single snap. I had no idea where he got it, or what drew him to this garment, or if it was even meant to be worn around the waist, but whenever Bill wore his towel it was difficult to be around him. The towel was very short, and when he slouched on the couch his balls would hang out. Whether this was appropriate or not was not subject to debate. Whenever I complained, Bill compared me to Joseph Stalin.

"Kirk!" said Bill once I was fully in view. He then went into a familiar routine about how Kirk liked to "get it on" with intergalactic women, who were, for some strange reason, always wearing sixties-style plastic miniskirts and go-go boots even though they were light-years from Earth.

"Light-years!" he said.

"Bill, it's late," I said. "You need to go to bed."

Our conflicts no longer had a slow build; now they started as if we had already been arguing for days, which in a sense we had, and now that I was taking a term off from college, ostensibly to help them out, I didn't feel obligated to be nice anymore. My mom was staying home with my half brother, and as I was turning over my paychecks from my telemarketing job, I had adopted the same wage-earner's entitlement that Bill had lorded over me when I was in high school.

"I need to finish my ice cream," Bill said.

On the way to the living room I had seen a half-gallon of Breyer's Fudge Ripple, empty and much abused, leaking on the kitchen counter. I looked at Bill's bowl. Even though most of it had melted, there was still a peak of ice cream towering above the rim. In fact, there was so much ice cream it made one wonder why he didn't make it easier on everyone—himself, the bowl, me—and work straight out of the carton.

I put my hands on my hips and stared. I stood there, hovering like a hall monitor, but Bill didn't see me because he was nodding off. A finger of ice cream spilled onto his T-shirt, which woke him up. Bill wiped it off with the heel of his hand.

"Whoopsie do," he said, another old family expression, this time a reference to an episode of *Alfred Hitchcock Presents*.

Bill wasn't high. I knew what high looked like now. Bill had given me a sample vial of methadone a few weeks prior so I could experience the drug. It was an interesting buzz, slow and steady and even, and I remember doddering around the house, both tired and alert. Catching myself in the mirror, I had studied my slack face and glassy eyes for future comparison and Bill looked nothing like this now. He was tired and drowsy, but underneath there was a hollow energy that I had never seen before.

"You're falling asleep before you can eat the ice cream," I said. "Which means you're not eating the ice cream you're supposedly staying awake to eat. Which means you should go to bed."

"Ah, you are a wise and noble man, lad," Bill said. He started slapping his face.

"Quit fucking around," I said.

"What does this mean?" he said. "This 'fucking around'?" Now he was referring to an old *Star Trek* episode, one where Kirk schooled an alien female who was unfamiliar with the concept of love.

"It's time for bed."

"No," he said. "Damn it, lad, I'm going to sit right here and finish my ice cream!"

Now it was getting weird. It was as if Bill were stuck between two modes: the need to make pop-culture references and the need to eat ice cream. Nevertheless, I backed off. I was still afraid of Bill, still vulnerable to his temper. If he raised his voice loud enough, I was trained to let him have his way. Even if I fought back, I could never match his verbal ferocity, and I didn't want a raging lecture about my inferiority.

Bill moved off the couch and sat cross-legged on the floor. "See?" he said, making a show of his stability. "I can do this," he said, holding out his arms, the ice cream bowl in one hand as if it were ballast.

"I can also do *this*," he said. "*Wheee!*" He tilted on his back and kicked his legs in the air, all while holding the ice cream bowl. He rocked back and forth, switching from person to overturned turtle, kicking his legs all the way.

"What the fuck are you doing?" I said. "Stop it!"

"Aye, Captain!" This was too much. "Aye, Captain" was another part of our family's secret language, only it was reserved for outsiders and it was an insult. Bill would say "Aye, Captain" to flight attendants or movie ushers or other line minders or quiet keepers who tried to get my parents to behave themselves. This phrase was always delivered with such sincerity and warmth that even when the recipient knew he was being brushed off, there was such friendliness that there was no way to respond.

"That's it," I said. "You're coming with me."

I led Bill and the ice cream into the kitchen. Saving face, he said he would eat it later, when he was more awake. Bill then pretended that he heard something going on in the living room, a burglar sound that needed to be investigated, but when we got there he admitted it was a trick to get back to the TV. If Bill were anyone else I probably would have called 911. The rocking on the ground, the irrationality, the strange willfulness, and the different "characters" he was doing—these are what a psychiatrist would call "symptoms." Instead I took Bill's actions as being deliberate, an act of defiance.

I had had enough. I grabbed Bill's arm and tried to pull him off the couch. "You're going to bed right the fuck now," I said.

Bill made himself heavy like a toddler, and then started clawing at the hand on his arm. I was surprised at how weak his effort was. We're both skinny men and about the same height. There were even people who would see pictures of us and think we looked alike. "It's incredible," Bill would say. "You're not even mine." We had never had any physical contact like this, had never played a manly game of basketball or wrestled for fun when I was little. I had always assumed that he was stronger because of the strength of his voice and the force of his will, but he couldn't even get one of my fingers off his arm.

Then he started howling like a wolf. He kicked his legs and made himself heavy and howled, his mouth forming an O that stretched toward the ceiling. At first I thought he was acting. In his theater days, Bill liked playing the villain, hamming it up for the audience, basking in their hate. His howl seemed so fake it took me a minute to see he wasn't kidding.

I knew because of his eyes, which were wet like a dog's eyes. Now I was scared. Bill squirmed out of my grasp and started toward the front door. I hopped off the couch and put my body between him and the door. This was when he got very reasonable.

"Lad," he said. "I simply care to go for a stroll. Around the neighborhood, to clear my head. Please get out of my way."

"Bill?" said my mom. She was standing in the doorway between the kitchen and the dining room. "What are you two doing?"

"Bill won't go to bed," I said. "He's acting weird."

Bill said he would go to bed in a minute. He was going to go for a walk first. My mom eyed us for a few moments. We looked back at her and then at each other. It was almost as if whatever was happening was simply between us. She went back to bed.

Once she was gone, Bill and I started in again, him wanting to go for a walk, me trying to restrain him. He pushed toward me, reaching for the doorknob, backing me up against the door. I was amazed at his willfulness, and at my own insistence that he stay put. I started laughing; this had to be a joke. Ever since I was five years old I had wanted Bill to leave us, to walk out on our family and embark on the zany adventure of his choice. Now that he was actually trying to leave, I couldn't believe how desperate I was to make him stay.

Then Bill stopped. He looked at his arms and ran his fingers across the back of his forearm. "I can see the molecules in my hand," he said. His voice was filled with fascination and surprise, like he had picked up a trombone and realized that he could play. I now saw that Bill wasn't Bill. He was someone or something else, and this time, when Bill lunged for the door, I hit him hard on the shoulder. I hadn't laid a hand on anyone since fifth grade, when, on separate occasions, I lost a fight with a next-door neighbor and hit a girl in my class with my *Kung Fu* lunchbox, which was less a fight than a flirtation, but I held on to Bill's upper arm and pounded on his shoulder like I lived by violence.

"Okay, okay," he said. "Stop. Stop."

I wouldn't recommend that hitting patients be taught in medical schools as a way of calming the mentally ill, but in this case a minute or two of punching seemed to be the answer. Bill said he was done trying to leave. "I think I'm going to lie down for a bit, lad," he said. There was a distant, broken quality to his voice. It was almost as if he was letting me down by going to bed, like he had been so much fun recently and he really hated to break up the party, but even the most interesting person in the world needed his rest, so what were you going to do?

I stayed up for the next hour watching *Star Trek*. My mom got up and came out to see how I was doing. I said that Bill had completely lost it and that we needed to do something. There was a baby in the house. Enough was enough.

The next day Bill was still pretty crazy, but he had at least stopped hallucinating. There were even moments when it seemed he might be willing to admit that he might possibly need to see his psychiatrist. I called Dr. M., who wanted to know if Bill had taken his medication. I told him that the day before, Bill had been counting his pills in an attempt to remember whether or not he had taken them, but that this was nothing out of the ordinary.

I called a cab. Bill said he thought going to see Dr. M. was for the best, but when the cab arrived he resisted. I forced him down the stairs. When he saw the cab, he was delighted.

"That's one of the great things about New York," he said. "You don't need a car. You can take a cab just about anywhere. And if you want to go away for the weekend, you can always rent a car."

When we arrived at Dr. M.'s, I pulled him up the stairs and the rest is the rest. Of course I ran after him; I'm a good guy. It wasn't hard to catch him, either. When I did, he seemed glad to see me and was pretty easily coaxed into a new cab. But Bill didn't make it to the mental institution, and I suspect that Dr. M. knew that he never would. As it turned out, Bill's physical state prevented his admission. He had a very high fever, a hundred-and-four-degree brain-boiler that required medical attention first. That night at the hospital Bill's stomach was pumped; he was given an IV to counter his dehydration and put in a room for observation. The next day his vital signs had returned to normal and he was lucid enough that in the eyes of the hospital staff involuntary commitment was deemed unnecessary.

Or at least that was how Bill's situation was presented to me, how his case was minimized by my mom. I never spoke to a physician myself, and when I went to visit Bill at the hospital my parents acted as if life couldn't be finer.

"Lad!" he said. "I'm glad you could make it."

Bill was in a great mood. He told the story of how he had struggled free and run naked down the hallway while a pack of nurses and orderlies chased after him. "Apparently I was quite a handful," he said. He spoke with an amused detachment, as if this had happened to a Dickens character. "I remember the nurse trying to pin me down. 'Mr. Cass,' she said. 'Mr. Cass!' Like I was a misbehaving schoolboy! Delightful!"

I felt bad, partly for trying to commit him and partly because I had been so unsuccessful. What did it take to get rid of a crazy person in this town? Neither of my parents made reference to my efforts. It was as if I had no part in why Bill was here, other than as a well-wisher. He said he didn't remember anything. He didn't remember Amsterdam Avenue, or howling like a wolf, or me punching him in the arm.

"So that's it?" I said. "You're okay?"

Bill became serious. "I need to be more careful about the Parnate," he said. "I realize now that I had taken at least three times as much as I was supposed to. I have no idea how I did that, but, obviously . . ."

"You said you could see the molecules in your hand," I said.

Bill considered this. "Well, I suppose I could."

Bill came home the next day. In anticipation of his return, I went to the video store and rented a handful of bad movies we could laugh at, and then to the grocery store to stock up on ice cream.

In my twenties, I was obsessed with why I froze when Bill got out of the cab and why I hit him. I used to think these were two of the most significant moments of my life. I always saw the former as a failure, a flaw in my character that kept me from acting, while the latter was a strange kind of triumph, me finally standing up to Bill.

But now both incidents seemed less me and more my brain, pure amygdaloid reactions that could have happened any time in our species' history. Fear researcher Joseph LeDoux talks about how the fight-or-flight response is really the *freeze*-fight-or-flight response. Rats that receive foot shocks in fear-conditioning experiments freeze on the spot—and often spontaneously urinate or defecate. That irritating horror-movie convention when the killer's intended victim doesn't simply run out of the house makes neurological sense. The first reaction to overpowering emotional stimuli is to be overpowered.

Any animal would have frozen getting out of the cab. As for hitting Bill, maybe there was some turning point there, but that's the future making sense of the past. At the time, I was scared and felt threatened. He had pushed me past whatever breaking point I have for violence. Rather than be proud—or ashamed, for that matter—I can take solace in the fact that if I am being assaulted by a hallucinating madman, I will fight back.

As I revisited that day twenty years later, what was more interesting to me was why it took me so long to figure out that Bill had gone mad. How had I misread him so badly?

I first looked for the answer to this question in something psychologists call Theory of Mind. The classic way to probe Theory of Mind is via a thought experiment called the Sally-Ann Task. Imagine a room containing a box, a basket, and a girl named Sally. Sally is a very lucky girl. She has candy. Sally takes her candy, puts it in the box, and then leaves the room. Ann enters. Ann wants to be good, but she's not the best person you'll ever meet. Ann takes Sally's candy from the box and puts it in the basket and then leaves. When Sally returns, where will she look for the candy?

Children under the age of four, as well as people who suffer from autism, will say the basket. Healthy adults and children age four and up will say the box. Toddlers and autistic people cannot separate what they know about the candy from what Sally doesn't know. Everyone else can imagine Sally's state of mind as she enters the room, blind to the change in her candy's location. Theory of Mind helps you understand that Sally harbors a false belief about the storage situation of her candy; it gives you the ability to walk a mile in someone else's brain. You can imagine what's going through Sally's head as she approaches the empty box, on the verge of disappointment.

Sometimes Theory of Mind gets sexed up and called "mind reading," but it's not nearly that interesting. Yes, you are "reading" someone else's mind, but the cues—eyes, facial expression, gestures, words—are open and public. Furthermore, the net effect of Theory of Mind usually isn't that dramatic—we do it all the time, sussing out when people are angry, sad, happy, or bemused with such regularity that the function is almost invisible. Theory of Mind is also far from foolproof. That sparkly stranger who is eyeing you across the room may be attracted to you, but she also might be a born-again Christian.

This, in effect, had been one of the chief problems in my relationship with Bill all along: Theory of Mind is only as good as the mind doing the theorizing. Looking back at the dinner-table conversation that signaled our move to New York, I could see the weakness of my mind's ability to read theirs.

Despite whatever disadvantage I had as a child in terms of not having voting rights in the family, even if I'd been fully vested I wasn't capable of talking my parents out of the move, because I couldn't truly understand their reasons. I had the tools, but not the experience. In the end maybe this was the story of New York.

"Any normal eleven-year-old has a deep understanding of mental life," Yale's Paul Bloom once told me after I had related the story of our move to New York. "You would be able to reason about desires, goals, intentions, true beliefs, false beliefs, but you would still be limited in a couple of ways. First, there are countless subtleties about how people think and behave that need to be learned. For instance, you probably had a limited understanding about how people think and behave when it comes to romance and sex. And second, you might not be as good as an adult at complex mental attributions: X thinks that Y suspects that Z worries that. . . . Mastery at this sort of Machiavellian intelligence takes many years to develop, and a typical eleven-year-old would be at a disadvantage relative to an adult."

There was another flaw in Theory of Mind, one that had to do with Bill's brain. I had based my reading of him on my own relatively calm, rational, and commonsense approach to life, when maybe the entire endeavor was born not out of Bill's mistaken notions about what he should accomplish in life—or where—but out of his mania. Maybe Theory of Mind falls apart if one of the minds involved is broken.

Manic depression might have been the key, but I still wasn't satisfied. In the fall of 2005, I went back to the science library at the University of Minnesota and pulled up more papers with the goal of atomizing the night Bill went mad. I whipped up a fresh batch of index cards dedicated solely to that night, and started moving them around on my office wall, depending on which brain structure or function or neurotransmitter or gene struck me at the time as being dominant.

First, I had to contend with the Parnate overdose, which, according to a medical website, can lead to "insomnia, restlessness and anxiety, progressing in severe cases to agitation, mental confusion and incoherence." I suppose Bill was agitated and mentally confused, but the Parnate came awfully late in the story. It was certainly a factor that night, but it didn't explain the underlying problem with Bill's brain.

Then I stumbled across an enzyme called protein kinase C (or PKC). Recent work by researchers at the NIMH has found that bipolar (a.k.a. manic-depressive) patients were vulnerable to excesses of PKC, which is triggered, under stress, by norepinephrine and leads to increased excitability in the prefrontal cortex. Too much PKC might have been responsible for the howling, and could have also led to what one researcher called "word salad." Maybe Bill didn't hallucinate at all, but was merely speaking a string of meaningless words.

I became convinced that the answer must be deeper than that. I moved more index cards around, dug up more papers. Attention . . . an amygdaloidal response . . . an imbalance in his neuroendocrine system—it was all happening, but in what proportion? Who was leading and who was following? Digging down into the next layer only made it worse. Serotonin, dopamine, norepinephrine, cortisol—each of these transmitters and neurohormones were in play, but, again, in what order of importance? I think this was just another way of avoiding the simpler truths about that night: if Bill's breakdown was something complicated, then it would be more forgivable. I tried wiring diagrams and charts and using Playmobile pirates to act out the chemical scene playing out in Bill's brain, but this quickly seemed like my own form of madness and I gave it up.

Simplification didn't work, either. There was clearly a lot of amygdala in the room that night, but the amygdala shows up everywhere anyway. We are emotional beings. Plus, whenever I would try to simplify, to construct a unified theory of Bill's madness, I always ended up dragging another system into play. I thought I had the answer with cortisol, which not only boosts norepinephrine and helps take the PFC offline, but also affects the HPA axis and the amygdala. But then that seemed wrong, because the amygdala needed to come into play earlier to trigger the cortisol in the first place. I also rejected addiction and reward because that felt too deterministic. Addicts had problems with self-control, but they were not automatons.

I ended up with a kind of compromise of all three, which I taped up on my office wall in a rough time line. First there was the base, the inappropriate serotonin level thanks to the depression, which led to overall brain dysfunction. Then came the compromised reward network, which led to decreased natural salience and further drug-seeking behavior, which then led to addiction, which created a homeostatic imbalance that was compounded by the external stressor of living in New York City in the 1980s, and being in finance when you didn't belong there. Social isolation and social stress from the ignorant masses also led to reduced levels in oxytocin, higher levels of which would have had a calming effect and provided a check on rising cortisol levels, which in the end must have been dangerously high, with concomitant fight-or-flight response potentiation and hippocampal dysfunction and immune-system weakening.

Add the sleep deprivation and oxytocin spike from having a

new baby and no wonder he was a mess. Then, suddenly, I enter the living room, where he is peacefully eating ice cream and watching *Star Trek,* and shout at him to go to bed. At first he doesn't see me as a threat, but as I continue to press, his HPA axis did exactly what it was supposed to do. Perhaps his bipolar disorder makes him more vulnerable, or maybe the stress of being solely responsible for the family's welfare pumps his cortisol levels up to the point where it takes less stimulus to set off his fight-or-flight response, or perhaps the Parnate has also elevated his norepinephrine levels. Maybe it's all three. In any case, his PFC shuts down and his brain starts running on pure amygdala, while his PKC levels further scramble the frontal lobes and he starts howling like a wolf and kicking on his back like a turtle and either having hallucinations outright or simply fixing up a word salad or two.

This analysis bore no resemblance to what I remembered experiencing that night, and it was entirely speculative—but it also felt close to right. This was what happened, this was the scientific story of Bill's brain the night he went mad and I tried to lock him up in a mental institution. It was absolutely right, I kept telling myself, until one day I looked at the wall of index cards and realized there was a crucial part of the brain I had left out.

The big hole in my effort—and it was a *big* hole—was memory. If the brain was any kind of machine, then it was a learning and memory machine. I had always thought of memory in terms of what neuroscientists call *autobiographical memory*, the explicit memory of events. But scientists also looked at memory in much broader terms. Part of my problems with attention could be seen as a deficit in short-term memory, or *working memory*, while the learning my neurons did during the fear-conditioning experiment could be considered a kind of *implicit memory*.

I first avoided investigating memory simply because I didn't like memory experiments, which were some of the most boring you can do. I felt like a memory experiment should involve putting on some kind of play that the subject didn't know about, like in a movie about con artists who have set up a fake investment firm, but it's all reception area and no back room. Then the subject is told later, "Surprise, that party you went to was filled with actors. Now, what do you remember?"

Well, forget that. The stimuli ended up being things like word lists and clip-art pictures and computer-led quizzes. No chocolate, no impromptu speeches about how much you hate yourself, no frantic space-bar tapping. Worse, the majority of memory research involved the animal model. Many big pronouncements about how memory works were based on a rat "remembering" how to run a maze.

But I think ultimately the problem was with me and not science. I didn't want to study memory because I couldn't handle any more failure and disappointment. My brain felt vulnerable and exposed, and I was afraid of learning something about memory that would further compromise my sense of self. Even though memory was on my original shopping list for my mind, I now felt that if I expanded my inquiry into Bill's brain and my brain to include memory, everything I had done until now might unravel. Science had done enough damage already. Even though I knew it was cowardly, and perhaps even disingenuous, to ignore memory, I decided to leave it alone. I decided that I had already remembered enough.

THE ENLIGHTENED BRAIN

Very early on in my research, I had chatted with Alarik Are-nander, then director of the Brain Research Institute at the Maharishi University of Management in Fairfield, Iowa. Are-nander was one of the first researchers I spoke to, and it had been fun to listen to him talk about the brain science of Transcendental Meditation (TM), "the intelligence of the universe," and "unbounded consciousness." We also talked about how his organization was looking at the neurological underpinnings of the "transcendent experience of the brain" and how he had "the first empirical evidence of the something called Enlightenment."

I had been saving my visit to Fairfield in the hope that my eventual trip down there would be a kind of hero's welcome. After battling with the brain, I had hoped to end on a transcendent note, with tangible proof of my own magnificence. By this point I was supposed to be stronger and smarter, but in fact I was beat-up and numb, living with a secret Adderall habit, smoking, and nursing an incipient midlife crisis. I felt like the brain had undermined everything that I thought about how the world and the self worked. I could see the quest to understand my brain taking over my life—ever probing and questioning and guessing at the meaning of it all—and as I was getting ready to go down to Iowa I felt a sense of doom. Studying my brain's highest functions seemed not only presumptuous but unwise.

Typically, scientists are smart when it comes to studying the ineffable—they avoid it. In his paper "Consciousness," John Searle, a professor of philosophy at Berkeley, takes the neuroscience community to task for this dodge, essentially calling researchers chicken. He notes that one of the main reasons scientists have avoided consciousness in particular is because it's a hard subject to tackle scientifically. At the same time, Searle argues that consciousness can, and should, be studied despite its squishy nature. Even if consciousness feels like an illusion, something manufactures that illusion, which means there is a mechanism that can be explored.

I felt like I had touched too many mechanisms already. There was a science-fiction story I recalled from my Seattle youth, perhaps by Isaac Asimov, where a supercomputer had calculated the scientific basis for humor. At the very moment the computer spit out the answer, everyone in the world lost the ability to laugh. I would hate to crack open the code for my consciousness only to ruin it for everyone else.

The work at the Brain Research Institute focuses on the brains of people who practice TM. I would be working with Fred Travis, director of the Center for Brain, Consciousness, and Cognition. Travis had been practicing TM for thirty-three years, and he sounded pretty blissful on the phone. In our preliminary talks he wanted me to know that all mediation and prayer was not created equal. "Christian prayer involves effort," he said, noting that the thalamus, a brain structure that plays a major role in our sensory awareness and general arousal, was still active. "Buddhists had thalamic activity, too. They involve effort. They do not give the experience of unboundedness."

With people who practice TM, however, although the attentional system is awake, "there is silence in the thalamus," Travis said. "It's the picture of restful alertness. There is a sense of awareness, but at a very quiet level. This very much supports the experience in TM of being outside of time and outside of space. You don't know what time of day it is. You don't know where you are. You don't know your own name. You're just awake."

My pessimism about the brain had gotten so bad that the idea of learning more about being "awake" sounded ominous. Still, I made the trip, mostly out of a sense of obligation. Like a marathon runner at mile 22 finding himself with bloody nipples and shit running down his leg, I felt that at this point I might as well finish it.

The morning of my visit to Fairfield, I kept myself clean—no cigarettes, no Adderall, no caffeine in my coffee, and fruit with yogurt for breakfast. As I approached Fairfield, the beautiful southern Iowa countryside didn't feel very salient, but given how I had been living, I doubt I'd ever pinpoint what was responsible for the decrease in my dopamine D2 receptivity. I missed my Adderall. The problem with my medication now was that if I didn't take it, the world felt extra flat. It wasn't exactly a hangover. There weren't cravings, but I noticed that the planet didn't have the sheen it once had. My thoughts felt slow and my vision dull. This was not the best day for a spiritual awakening.

When I arrived at the Maharishi University of Management I was surprised it wasn't more resplendent. The Maharishi had been doing his thing since the sixties, when he introduced Transcendental Meditation to the world and acted as guru to the Beatles and the Rolling Stones. As for what exactly TM was, I wasn't sure. Some said a religion, others accused it of being a cult, still others a simple, natural, almost effortless way to achieve peace and bliss. Either way, the Maharishi now had big ambitions for the world, dreams of meditation centers and Eden-like cities and an end to all war, which was why I was surprised at how shabby the campus looked. I drove around, easing past the Maharishi Patanjali Golden Dome and a handful of dorms that looked like army barracks, and then around a series of streets with names like Taste of Utopia, and Golden Lane, that delineated a trailer park that looked like a taste of something else entirely.

It was early in the morning and the campus hadn't quite woken up yet. There were a few students walking around, the boys in khakis, dress shirts, and ties, the girls also conservatively dressed. I stopped a young woman and asked her where the student center was. She stared at me for a bit and said there really wasn't one, even though I later learned that we were standing right in front of it. One of the notions of TM was that practitioners developed a sense of "unboundedness," a pleasant blurring between themselves and the world. Indeed, people started saying hello to me as if I were a happy extension of themselves, and there was a sunny, warm, slightly spacey air of peace and calm about the entire campus. When I parked my car, I locked the door.

Then again, I was primed for this to be weird. Bill raised me to be a staunch atheist, and I considered it a point of great pride that I had blossomed into a hopeful agnostic. In fact, in the past few years I had visited mediation centers and had taken yoga classes for an article I wrote for a men's magazine. The whole idea of dedicating your life to spirituality seemed suspect to me—and my friends who grew up in Iowa gave me some very knowing looks when I told them about this field trip—but I was willing to keep an open mind, if not for God, then for Science.

I met Fred Travis on another part of campus that contained newer buildings—crisp yellow-and-white colonial-style houses that reminded me of pictures of Celebration, Florida. Fred was dressed in khakis, a white dress shirt, and Birkenstocks. He was medium height, with a shiny-bald dome head and softy soft brown eyes that had a small square of light in each pupil, as if he were a Japanese anime character.

His lab, which was housed in the Maharishi Veda Bhavan Building, was airy and fresh, with sunshine lightening up sheer curtains that waved in the Iowa breeze. There were the requisite posters of the brain on the wall, and the mandatory bit of brain whimsy, in this case a plastic brain wearing a pair of eyeglasses.

Travis installed me at the computer and started prepping me for an experiment. He measured my head (56.5 cm) and fitted my scalp with a 32-electrode-array cap (size: medium). He then hooked up the cap to an EEG machine. The star of the show today was my folds.

Travis explained that the tasks I would do today were the same that first-year Maharishi University of Management students take during orientation week. MUM was probably the only school in America that highly encouraged daily meditation, and Travis said it was important for the students to see their progress. They will take the test again in their second year and before they graduate. Travis called it a "brain report card."

Task one was a simple reaction-time experiment: when I heard a tone I clicked a button to make it stop. Task two was a simple choice game. I would be presented with one number and then another. If the second number was larger than the first, I was to click the button in my right hand; if the first number was larger, I was to click the button in my left hand. The final task was the moneymaker. It was another Conners product and similar to my attention test. A string of two hundred letters was presented, one right after the next, in rapid succession. If a letter was different from the previous letter, I was to press the left button. If a letter repeated, I was to press the right button.

I got situated, taking the response buttons (which were fashioned out of BMX handlebar grips), and settled in for my trials. I was far from experimentally naïve, especially where prefrontal matters were concerned, so I didn't sweat this one. The stimuli were presented. I pressed the correct button or I didn't. Subsequent stimuli were presented. I wasn't trying to beat the machine or prove anything this time. I knew that it wouldn't matter if I did. The EEG read me uncritically, and the data would show what the data would show. If I was enlightened, I was enlightened; if I wasn't, I wasn't. I told myself that I wasn't going to get mad either way.

After the trials were done, Travis needed to take a conference call. I waited out in the lobby and started to peruse some of the posters hanging on the wall. I had seen these on my way in but hadn't paid them much mind—they had seemed like abstract art inspired by computer chips. Then I read the text and learned that the Maharishi wanted to raise billions of dollars to reconfigure every city on the planet to these peaceful grid patterns. There was also a poster about a TM meeting in Washington, D.C., in the nineties that claimed that because of their mass meditation, crime went down and Clinton was finally able to pass more peace-oriented legislation. Until then I hadn't realized TM could be so ridiculous.

I went into the bathroom to clean up. I had conductance paste in my hair and some encrusted on my ear. I experienced a similar moment of clarity as I had in the scanner in Lehericy's lab, only this time when I thought, *What am I doing here?* the emotional value was one of dread and shame. I thought, *What kind of animal does this to itself?* Here I was, covered in science cum and waiting for a bunch of strangers and their machine and their weird beliefs to tell me who I was.

Washing up, I also started thinking about the hubris of this entire endeavor, and it seemed that Bill was the cause of this again. All this striving and struggling to prove how smart I was, to learn the unknowable, to do the impossible and comprehend the brain, when I was much happier essentially hiding in Minnesota. If I had left New York behind, why was I still trying to prove myself? Whose ambition was this, anyway? Was self-knowledge really worth all this humiliation?

I toyed with leaving; it would have been easy enough to simply get my data later, but I went back in to see Travis when he was done with his call. I couldn't tell from his demeanor how I did. Where sometimes I felt like my face was a series of contortions, scrunching to understand, frowning at something I didn't like, spreading open in delight, Travis had the same warm, friendly expression the entire time. It would be tempting to say it was childlike, but it wasn't that pure. It wasn't the complete absence of guile, as I sometimes saw in Owen, but an extra layer, a kind of veil of disengagement that neither took from you nor gave you anything. I guess it's hard to use Theory of Mind on someone who was unbounded.

First I needed a tutorial in reading an EEG. There was a lot to absorb, about how the lines rose when neurons were about to fire, then fell away after they had discharged, not to mention the different bands of waves—alpha, beta, and gamma—cycling at various numbers of hertz per second. The computer screen

Travis brought up had a lot more squiggly lines than I could ever make sense of, row after row descending from the top of the screen, some parallel, some intersecting with each other. It made about as much sense as a child's scribbles, so Travis added some filters, cutting down some of the "noise" and focusing on one of my alpha waves.

As with the galvanic skin response, I could see the moments of rest and the moments when I was responding to the task. There were even big effects—the whole line plummeted to the bottom of the page, then came sharply up—when I blinked my eyes. When I was at rest, the lines were high and close together; when I was doing the task, they were lower and flatter. This reflected the nature of alpha waves, which are inversely associated with effort. When you're at rest your alpha waves are high, when you're stimulated they drop.

Travis said that the interesting thing about TM and alpha waves does not happen during meditation. "Someone meditating for four months looks the same as someone who has been doing it for four years," he said. The significant change was what happens during the task. "You see more frontal coherence during whatever you're doing," he said. "There is more alpha power and less gamma power. You appear to be less concerned with surface perceptual differences and more with the underlying wholeness." People who mastered TM ended up keeping the same high levels of alpha waves no matter what the stimuli is. They were at rest even when they were being stimulated. Meanwhile, people who didn't practice TM were "like a football to the environment."

"You have a lot of alpha waves in the front, which is unusual in people who don't meditate," he said, pointing out a section of the graph. "Your mind is quiet, rested, which is a marker of what grows with meditation practice."

At first I took this to mean that I was indeed a higher being. Okay, now we're talking. But then I started staring at the low, flat part of the line that followed immediately after the praiseworthy part. Each of the alpha lines on the EEG followed this pattern. Squiggle, squiggle, squiggle before the stimuli was presented, then low and rolling during the response.

"I'm not," I said. "I'm not, because if I were then I would have those waves during the task."

Fred nodded, pleased. I understood.

I left Travis's office feeling as stupid as when I had arrived. On the way to the car I started arguing in my head with Travis about what his research showed. Even accepting the TM definition of enlightenment, I wasn't sure unboundedness was the point of the brain. For one, it was too prefrontal, too top down. This didn't make it not true—I had no doubt that Travis and his friends were a blissful bunch—but where was the reward network? Where was the HPA axis? Wasn't part of the whole point to be a football with the environment? To be engaged in the moment fighting for what you want?

When I realized I was standing in the middle of rural Iowa having imaginary discussions with a scientist, I had to stop myself. This whole endeavor had gone too far. I realized that this could go on forever. I would never get the answers I wanted.

What had started with so much pleasure—that absolutely wonderful insight feeling—had become almost relentlessly painful. In his book *Mind Wide Open*, science writer Stephen Johnson coined the term "recreational neuroscience." Well, I, for one, was not having a good time. That researcher who had warned me against brain self-awareness was right—I didn't want to see the glasses. I wanted to be myself again, even if that self now seemed besides the point, one of the many systems that were designed to keep me alive and reproducing, but little else. Thought, memory, consciousness, self—everything we celebrate as being distinctly human seemed like mere survival tools, and survival satisfies its needs at a depressingly low threshold.

I took a breath and looked around. It was a beautiful day. I was on a college campus, which is rarely a bad thing. Then, to my relief, I saw a rocker chick wearing a black T-shirt with the name of a heavy metal band written across the front in silver glitter. Amidst all the conservative-looking kids she would be my rescue.

"Do you have a cigarette?" I said.

"No." She sounded like she felt sorry for me for even asking.

"Probably not anyone who smokes on this campus, huh?" I said.

"Probably not," she said.

It was so typical of how the entire endeavor had gone. To be on a college campus and not be able to bum a smoke seemed cruel, but I also took it as further proof that I did not belong here and that there was only one way to put an end to the spiraling madness of the brain.

The spring term of my sophomore year I simply stopped taking my antidepressants. Then, instead of going home, I went to Seattle for the summer and painted houses and hung out with childhood friends, and after that I only became more and more estranged from my parents. I stopped visiting, stopped going home for Christmas, then stopped calling. In my angry twenties I thought that if my parents cared about me they would call me or write to me, and they didn't call me, so I didn't call them.

There was a period in the mid-nineties when I didn't talk to them for an entire year, and in that time their situation only got worse and worse. They both lost their jobs, went on disability for depression, drank, abused drugs, defrauded pharmacies, and God knows what else. By the time I got married, they had suffered so much and for so long, I hardly knew them.

Quitting my family was one of the most difficult and painful experiences of my life, but it worked. I survived, which I suppose is all a brain wants; and if that was the case, then survival was what my brain would get. It was at that moment down in Fairfield, Iowa, that I told myself, *Enough*. I am not my stepfather, and any endeavor that has me thinking that I am must stop. No more science. No more self-discovery. From this moment forward I would embrace ignorance and sloth and self-deception. That's it, brain. I am done with you. I quit.

Two years ago, in the fall of 2004, I attended the Society for Neuroscience's annual meeting in San Diego for what I thought was going to be my farewell to science. Before my experiments with Adderall, and my foray into the genetics of my dopamine receptors, and my trip to Fairfield, Iowa, I had gone looking for an ending, some final idea that would sum up my adventures.

I felt pretty good. For one, you cannot be depressed in the San Diego Convention Center. Located on the harbor, the convention center has a nautical theme, with porthole-like windows and ushers dressed like ship's stewards in long white jackets with gold-braided epaulets and jaunty sea caps. Even though it was filled with scientists, the whole place felt like a magic sunshine cruise ship. Come aboard—we're expecting you.

As usual I had no plan, but the idea—always ideas, never plans—was to explore the minutiae of the brain. I was going to go where I had previously been afraid to venture, strange, exotic places like Purkinje cells and ion channels and phosphorylation, which admittedly, isn't a place but a process.

In a section innocuously called Development, I approached David Sandeman, a biologist from Wellesley College and the primary investigator on a poster presentation titled "Neuronal Proliferation in Adult Crayfish Brain: Effects of CNS Ablations, Molt Cycle and Temperature."

Sandeman, a middle-aged Brit in silver-framed glasses, a blue work shirt, and pale khakis, spoke with such brightness, intelligence, and warmth you would think he was hired to play the kind of movie college professor who connects with his students and changes their lives.

Sandeman explained that the dogma used to be that once you hit a certain age you had a fixed number of neurons, which then slowly but surely started to die off. This was no longer thought to be true. Now scientists know that certain parts of the brain are capable of growing new neurons even in advancing age.

"This is enormously significant for people in the medical field," he said. "Maybe we can get neurons to grow. It's not quite as simple as that, but there is also a clear relationship between activity, learning, and the addition of neurons. This idea of adult neurogenesis has blown the whole field wide open. Now we want to know what switches a cell cycle on, what turns it off."

This sounded really good to me, life being better than death. It sounded good to Sandeman, too. He became only more energetic and passionate.

"The take-home message is the following. The brain is an organ like all the rest. Like muscles. You don't use it, you lose it. It's absolutely *essential* you use it. There are so many cases of people who go on being active, who remain intellectually challenged. They don't suffer from degenerative diseases. No, that's not quite true. That's taking it too far."

"Because then you could just say that dumb people are screwed," I offered.

"No, no, no," he said. "Some kinds of senility can be avoided if . . . you . . . keep . . . the . . . brain . . . busy. Normal aging can be a lot more comfortable if you keep the brain going. You can see it in the elderly who play bridge and keep themselves active. They're bright as buttons!"

I looked over at his poster, which, in addition to the usual graphs and charts, featured pictures of lobsters and crayfish.

"Even in lobsters," I said.

"Even in lobsters," he said.

He then explained his experiment, in which some lobsters were isolated and confined, while others were allowed to roam more freely and interact. The lobsters that were confined showed a drop in neurogenesis, which supports research that has shown similar declines in patients suffering from depression. The question for Sandeman was to see if the sad, brain-decaying lobsters could recover. So he gave the lonely lobsters some company.

"One of the interesting things is that it would seem that the actual process of establishing dominance hierarchies is what makes the difference," he said. The lobsters, as you would imagine lobsters would, fought for control of the tank. Given all the conflict I grew up with, not to mention the studies I had read about the social stress experienced by olive monkeys, I assumed that fighting would be detrimental to the losing lobster, but Sandeman said that "as long as there is competition going, then we see neurogenesis."

"Even in the submissive one?" I said.

"Yes," he said.

"So you don't necessarily have to win," I said.

Sandeman was quite pleased that I had arrived at this conclusion. Indeed, it was the process; it was the fight that mattered. This was what I had been looking for. That even in my drug-addicted, mentally-ill stepfather, the brain was a place of growth and renewal. More important, all the misery he had put me through had perhaps even made me stronger. The message was to keep fighting. Stay out of that restaurant fish tank and keep the rubber bands off your claws. If you do, your brain will blossom and bloom, and even if the changes aren't dramatic or miraculous, they're positive change nonetheless. That was it. I truly had found the answer this time, and a happy ending to boot.

"You know I married your mother in spite of you and not because of you," Bill said to me over the phone. "You know that, don't you?"

Owen was at day care and Liz was at work, and I had been reading through a paper on reward when Bill called. The fact that he was making the effort told me that he was probably in between one of his swings, one of those rare moods when he was neither too depressed to pick up the phone nor so manic that he knew I wouldn't want to talk to him. Still, I continued to surf the Internet, in an effort to divide my attention. With all the love in my life, the sound of Bill's voice could still put me in a funk for weeks, especially at times like these when he would binge on the truth.

Ten years ago I would have fought him on his comment. I would have railed against his insensitivity and lack of respect. I would have defended my life in Minneapolis, how I was happy and had friends and liked my career, and he would counter that it was a piddling place and unworthy and an indication of my mediocrity. Or I would have gone on the attack in a vain effort to make him understand how New York and his drug addiction and even his personality had been destructive forces for our family.

Now I just said, "I know."

He was older now, living on disability, and almost completely broken, but we had also come to an understanding. He was not to yell at me or lecture me, he was not to interrupt me or get mad when I told him that he had been talking for twenty straight minutes. These are small victories. What a story like ours desperately needed was a full reconciliation.

For my part I had forgiven Bill many times, but now that I was almost done with the brain, I understood that too much was needed between us for there ever to be true peace. First, he would have to get well, and then see the error of his ways, then overcome his inherent personality problems and show atonement through actual deeds. Then I would have to resolve years of anger, and continue to forgive his inevitable bad behavior—even if he got better, he would still be him—then be receptive to Bill's overtures, and probably help guide him in our new relationship. All this seemed like a lot to ask of two prehistoric minds. I'm not sure even the most dedicated lobsters could pull off that much renewal.

There were times when I felt that Bill was a brain and other times when I felt that he was a person. "I really didn't think much of you until you were about six," he said during another phone call and he sounded, as he often did in moments like these, like this was fresh news I should be grateful to hear. I tried, but I couldn't assign this behavior to some malfunctioning anatomical structure or system. When Bill talked to me this way he wasn't a brain; he was a shitty dad.

There were other times in those last phone calls when he was so addled and damaged that he seemed very much like a brain, delicate and gelatinous and incapable of supporting its own weight. What if he had some kind of mutation for the tryptophan hydroxylase-2 gene and was thus physiologically doomed to have only marginal success with antidepressants? Or what if the damage he had done to his hippocampus over the years had made it almost impossible to down-regulate his stress response? When I was feeling sorry for Bill, rather than angry at him, he became the sum of the neurological strikes against him. Of course he acted this way. How could he not?

As for the story of our time in New York, the way I saw it, Bill was completely off the hook for the initial self-delusion. I had made such a mess of my own brain research, how could I fault him for thinking New York would make him, or for the fact that the city eventually ground him down? At the same time, I held him more accountable overall. We're often not responsible for our mistakes, only for trying to correct them. If a place or a job or a person is destroying your physical and mental health, then *you leave*.

I never got the chance to tell Bill about these findings. For the three years I investigated my brain, both he and my mom were completely ignorant of my research. Then, on New Year's Eve 2005, Bill died in the middle of the night while sitting in his easy chair. He was fifty-six years old. According to my mom, the county coroner later found methadone in his blood, though she said it was unclear whether or not this was the direct cause of death.

The methadone revelation was a surprise but not a shock. I learned that my mom had been taking the drug again for her back pain, and that Bill had stolen an indeterminate number of pills from the various caches my mom had hidden around their apartment in Seattle. Strangely, the fact that either of them were taking—or even abusing—prescription drugs didn't bother me, nor did I feel an overwhelming sense of grief at his passing. What bothered me was the fact that the coroner's report lacked finality. Even in death, Bill—like science—left me with still more uncertainty.

• • •

The decision I made down in Fairfield to cast the brain out of my life has proven partially successful. Some days I wear a light, extra layer of brain awareness; other days my brain is a drag. Sometimes the knowledge intrudes, like when I'm spending time with Liz and Owen and silently pin their behavior to their amygdalas. Other times, like when I'm arguing with my stepfather in my head, the knowledge helps me turn my brain around. I can tell myself those thoughts are mental manifestations of stress or anxiety—not me, but my body.

I take my body a little more seriously now. When I got down to my last dozen or so Adderall, I debated whether or not to finish the prescription. Do I *want* to take a time-released, low-dose amphetamine? Or are there other ways of boosting saliency? Do I accept the cognitive declines that come with being a social drinker or do I change my lifestyle to better my brain? Eventually I flushed the rest of the Adderall down the toilet. I was hoping for a dramatic, conclusive ceremony, but as I watched the tablets dissolve in a blue mist, I could only hope this was the last time. As for smoking . . .

If only my brain would sit still. Then maybe I could end this entire adventure with some final message, some bright bit of wonder and uplift. All the other popular neuroscience books end this way. After coldly dismantling the brain's structures and processes, the authors invariably feel the need to warm up the ending with a little old-timey humanity, some note of hope or redemption or a sly question that points to future understanding. I can't do that. The person in love with insight back in 2002 could have pulled that off, but that man seems like a stranger now. I can't lay it on too heavy, either—I've dispensed enough evolutionary doom—nor do I feel right offering up a Zen soft-ball (the brain simply *is*) or a public-service announcement that calls for everyone to treat their brains to plenty of exercise, sleep, and good, fresh, nutritious foods.

One of Owen's Baby Einstein videos opens with a quote attributed to Galileo: "All truths are understandable once they are discovered. The point is to go out and discover them." Maybe there will be a final, uplifting truth about the brain. Maybe some Christ-like man-child will come along and not only solve the brain, but also make the solution play well to the masses, ignorant and otherwise. I have my doubts. The real truth about the brain may not be understandable, and I'm not sure if anyone wants it to be discovered, anyway. If science ruined God for everyone, neuroscience risks ruining the Self. Even if science finds something big and flashy and incontrovertible, there will still be people who will reject this truth, just to be that way.

The biggest question remaining is: What am I going to tell Owen? You can't tell a three-year-old that right this instant he's training his brain for future life, and that someday he may want to change or be different but it's going to be too late. Even if, for example, I could impress upon him the importance of keeping his reward network pure it wouldn't be the point. (Maybe that was what Bill was trying to do with his readings lists and his lectures. How funny it would be for him to see me now, with my love of reality TV.) Whatever I might try to do to educate Owen about the science of himself would be futile. Even with full brain control there would be too much to manage. I promise to do everything I can for him, to love him and support him, but when it comes to his brain, he's on his own.

AUTHOR'S NOTE

Head Case is a work of nonfiction. Some of the dialogue has been edited for clarity, and a handful of scenes appear out of chronological order. The scientific theories, experiments, and concepts are all factual. In translating and applying the science, I have done my best to be as faithful to my source material as possible, but as these pages show, I am only human. As a result, the information in this book is to be used for entertainment purposes only.

ACKNOWLEDGMENTS

Thank you to Alison Callahan at HarperCollins for believing in this project from the beginning and for continuing to believe, even when it would have been well within your right to stop believing. Thank you also to Jeanette Perez.

Thank you to Heather Schroder at ICM for being a wonderful agent and a real person, and to Margot Meyers for all the pep talks.

Thank you to Clara Jeffery, Dara Moskowitz, and Eric Vrooman for their years of writing advice, support, and friendship.

Thank you to the readers: Jan Borene, Susie Kukkonen, Jill Locke, Kate Muhl, and Todd Pruzan.

Thank you to Lisa Erb for helping with the art.

Thank you to all the scientists in this book. A special thanks to Trent Jerde, Thomas Naselaris, David Santucci, and the other young neuroscientists who showed me the human side of their profession. Thanks also to the neuroscience community in general. The scientists I met while researching this book—even the one who yelled at me—deserve our utmost respect and admiration. Theirs is not an easy life, and yet the contributions they make, no matter how fraught with complications, are genuine and necessary.

Finally, thank you, Liz. Even when things got hard and weird, your love and support never wavered.

Insights,
Interviews
& More . . .

About the author

About the book

Read on

Meet Dennis Cass

Liz Barrere

DENNIS CASS is a writer, teacher, and occasional public speaker. His work has appeared in the *New York Times Magazine*, *GQ*, and on Slate.com. He lives with his wife and son in Minneapolis, Minnesota. Visit him on the Web at www.denniscass.com.

A Short Conversation Between Dennis Cass and His Brain

Dennis Cass: *Thanks for joining me. I know you're very busy and I appreciate the time.*

Dennis Cass's Brain: My pleasure.

DC: *The book is out. People are reading it. How do you feel about the way you were portrayed?*

DCB: I think it's mostly fair. It's weird to be in a book because you see things that look wrong, and your first reaction is, *That's not me!* But then you realize it's not wrong so much as it's the author's point of view. It kind of makes me feel sad for movie stars—all the crap they have to read about themselves.

DC: *Anything stand out as particularly unfair?*

DCB: I think there are times when I come off as having it out for you, which doesn't seem right. I mean, I've gotten you through a lot, and you seem to be doing fine. So it's less the specific things you got wrong and more the overall tone. I'm sitting there thinking, like, *Am I really that bad?*

DC: *What parts of the book did you like the most?*

DCB: I like this idea that the brain is a ▶

> *It's weird to be in a book because you see things that look wrong, and your first reaction is, That's not me! But then you realize it's not wrong so much as it's the author's point of view.*

3

moving target, that whenever you try to understand one part you're neglecting another. Because I'm so like that. I'm not going to sit still for pictures. I'm not going to make it easy for a guy like you to come along and figure me out.

DC: *I've thought about you, done research into you, and written about you, but still I don't quite understand how you work. What's your process?*

DCB: I don't have a set way of doing things. Generally, I take stuff in, understand it as best as I can, and then direct thoughts, feelings, and actions in a way that I hope is consistent enough with reality so as not to create a disconnect between the outside world and your internal sense of yourself. My motto: Keep it smooth.

DC: *Is this what you saw yourself doing back when you were younger?*

DCB: Yes and no. I knew that being a brain was what I was meant to do. And I'm grateful for that. That's such a gift. What I didn't realize was that so much of my existence was going to be the same. The *same* worries, the *same* fears, the *same* joys. Part of me is going, *Can't you come up with something new?* And then another part of me is going, *No, that's cool. Just keep being you. You've got enough on your plate.*

DC: *Because it's work.*

DCB: It is work. A lot of work. And constant work. *You* get to sleep. *You* get to zone out. *You* get to watch movies and play video games and get drunk. But what am I doing when you're doing all this stuff? Working my ass off. Working to make you sleep. Working to help you escape into a fantasy world of your own or someone else's devising. Working to be drunk. So there is no break for this guy. But it's what I do, so—

DC: *And you really don't have much choice in the matter.*

DCB: This is true. This is true. If I want to work, then this is the work I get to do. So you just try to make good choices and use your time wisely.

DC: *You mean I make good choices and use my time wisely? Or you make good choices and use your time wisely?*

DCB: There is a difference there, but not one that we're going to figure out today.

DC: *So what's next?*

DCB: I imagine a lot of that is up to you. But if I were calling the shots, I'd stay away from science.

DC: *Too hard.* ▶

> ❝ *You* get to sleep. *You* get to zone out. *You* get to watch movies and play video games and get drunk. But what am I doing when you're doing all this stuff? Working my ass off. ❞

A Short Conversation Between Dennis Cass and His Brain *(continued)*

DCB: Too hard. Too sad! And I'm not sure the answer is in the brain. At least not for me. And especially not for you. I think you should find something else to do. Like, maybe teaching.

DC: *You don't think I'm going to make it as a writer.*

DCB: [Mumbles.]

DC: *I didn't catch that.*

DCB: Look, man. You're great, and I love you. Can we just leave it at that?

DC: *You're the boss.* ∿

Head Case
Deleted Scene

At some point in the writing of this book, I came to the realization that, given the subject matter, Head Case *had the potential to run very, very long. So instead of writing the 2,700-page version that would be released in nine volumes over the course of my life, I decided to keep it short. The consequence of this choice was that I felt compelled to cut a number of scenes that I liked but that didn't fit, such as the following story of my first scientific conference. As much as this experience fits thematically with the rest of the book, every time I tried to bring this scene back into the manuscript, it felt like a detour. I probably added it back in and cut it out again half a dozen times, and even though it's appearing in an addendum, it's nice to have it back.*

THE NIGHT BEFORE I attend the 2003 Society for Neuroscience annual meeting, I write a note to myself in my notebook: *Attend as a peer.* I'm joking, but I'm not kidding. A year into my research for a book about my brain, I have developed this strange idea that, like an FBI profiler on the hunt for a serial killer, in order to understand science I need to learn to *think* like a scientist. I used to write about popular culture, television, and food, but that is all behind me. Now I am a man of science. *Attend as a peer* is my way of making sure that tomorrow I become someone else.

The sheer size of SFN, one of the ▶

world's largest scientific conferences, should be enough to convince me to keep my day job. The next morning, my stomach full of beignets and strong coffee, I walk through the door of New Orleans's Ernest N. Morial Convention Center, and what I see makes me queasy. The main hall of Morial is big enough to accommodate an air show, and it's packed with row after row of cloth-covered display stands that spread out in all directions and stretch to the horizon. Each stand is the size of a blackboard and is staffed by a neuroscientist. More than twenty-four thousand neuroscientists from around the world have come to New Orleans to exchange ideas, gossip, carve out or defend territory, angle for funding, find a job or at least keep the one they had. What I had thought was going to be as sleepy and genteel as a junior high science fair turns out to have the restless, desperate energy of a commodities market before the opening bell.

SFN has slide presentations, symposia, and public lectures, but I'm here for the posters. The poster is the dollar bill of science. Each measures about five feet wide by four feet tall and contains text, graphs, data, and images that together represent a year or two of work. Most of the researchers presenting posters are graduate students and post-docs. (The big guns take part in themed symposia; the giants, like National Institute of Drug Abuse director Nora Volkow, give public lectures.) There are roughly

66 What I had thought was going to be as sleepy and genteel as a junior-high science fair turns out to have the restless, desperate energy of a commodities market before the opening bell. 99

8

fifteen thousand posters floating around New Orleans—that is, fifteen thousand ideas that can give me a better sense of how my brain works.

I feel smarter just for being here, but then I start noticing some of this morning's sub-themes. Placards sticking up over the rows and sections of poster stands announce subjects like MONOAMINES, GABA; PROGENITORS AND PATTERNING; RECEPTOR TRAFFICKING/ SCAFFOLDING; and ION CHANNELS: VOLTAGE-GATED, INWARDLY RECTIFYING AND OTHERS. I'm looking for something a little more human, when I pass a poster titled PULLEY SHIFTS EXPLAIN VERTICAL DISCONJUNGANCY OF HORIZONTAL ANGULAR VESTIBULO-OCULAR REFLEX (AVOR) DURING CONVERGENCE AND VERTICAL GAZE. I avert my eyes and tell myself that as long as I can make it to the Cognition and Behavior section, I will be fine.

My first poster is called NEURONAL MECHANISMS FOR DECISION OPTIMIZATION. I picked it from the SFN catalog—which is so big it comes in five parts—because it sounded like the Holy Grail of physiology and psychology. The goal of my project is to connect the workings of the body (neuronal mechanisms) with my own particular mental concerns (as an expectant father, I would absolutely love it if my decisions were optimized). It's heartening to see that a handful of scientists also seem to be interested in this topic; apparently, my instincts are good. ▸

The presenter, Rafal Bogacz, from Princeton, is a scruffy Eastern European man in a flannel shirt. Bogacz is energetic, animated, and articulate, and I figure that when I reach the front of the line, a meeting of the minds is inevitable. If only his poster didn't look so completely alien and bizarre. First of all, it's far too busy. No "less is more" aesthetic here. I count twenty-two graphs and diagrams and fifteen subheads with titles like "Optimality of Diffusion," "Threshold Maximizing Reward Rate," and "Connectionist Model." And the equations are so complicated that not only can I not understand them but also they make my head hurt just to look at them. In one of them, $x_1 = (i_1 - i_2) \div \sqrt{2} + c$ is multiplied by what appears to be a chocolate chip cookie.

"We assume that subjects process decisions optimally," Bogacz says to me after I introduce myself. He then offers a bit of background about the ways human beings have evolved to optimize decision making.

"Like the eyes," he says, fluttering his hands in front of his face, "designed in this crazy way."

I write that down in my notebook: *Eye design = crazy.*

"Right," I say.

"The poster is in two parts," says Bogacz. "First, what is the optimal architecture of networking for making decisions? What is optimal speed versus accuracy? These neurons encode the

> 66 The presenter, Rafal Bogacz, from Princeton, is a scruffy Eastern European man in a flannel shirt. Bogacz is energetic, animated, and articulate, and I figure that when I reach the front of the line, a meeting of the minds is inevitable. 99

motion, the dark line above the light line, but the information is very noisy so we try to base the decision on MT neurons. What you can do is integrate over time—"

All this comes out incredibly fast. Bogacz talks so quickly it's as if the words are racing one another out of his mouth. Speed aside, I'm confused as to where we are in the brain. I thought we were in the whole brain or perhaps in populations of neurons, but it seems like we're in the individual neuron. If so, then how could a neuron "decide" anything?

Scientists trade in data, but when they are speaking to the press they are supposed to turn the data into stories. This is how a study about the antimutagenic qualities of an enzyme found in tomatoes ends up with the headline "Pizza Prevents Cancer?" Scientists are coached by their more experienced peers to talk to laypeople as if they were talking to someone's grandmother (who presumably is standing at the back porch in a gingham dress dusted with cake flour). But the way Bogacz is going on I imagine that his grandma runs the particle accelerator at Cal Tech. There is no story, no sound bite, not even a nod to the fact that I am a layperson. I wanted to attend as a peer, and I'm doing it.

"This is done with electrophysics?" I say. I throw out this term, which I picked up a few weeks ago, in an effort to show that I'm keeping up. ▶

> 66 Scientists trade in data, but when they are speaking to the press they are supposed to turn the data into stories. This is how a study about the anti-mutagenic qualities of an enzyme found in tomatoes ends up with the headline 'Pizza Prevents Cancer?' 99

Head Case (continued)

"Yes," said Bogacz. "These are single-unit recordings in monkeys."

Then I get the first of many very subtle looks from Bogacz, as if he thinks maybe I am pulling his leg. By bringing up electrophysics I have completely revealed my ignorance, my non-peer status. The *only* way to measure individual neurons is through electrophysics. It's as if Bogacz had been telling me about a Caribbean cruise, and I had asked him, "And did you ride on a boat?"

Then someone behind me jumps my turn.

"You're saying the integral here is essentially flat," the newcomer interjects, pointing at the poster. He is a stout, swarthy man who appears to be dressed for a day of hiking.

"This is biology," Bogacz says.

I think, *What* is biology? Or *whose* biology? With the exception of the comment about how the eyes are designed, you would not know we were even talking about the human brain, much less the people these brains inhabit.

"I don't see how you can integrate the difference if you're only seeing activity in one unit," says the camper scientist.

Bogacz explains how. I stand there like a cow. This is what it's like to be illiterate. You can feel the significance of words—after all, these squiggles and shapes are everywhere—but you have no access, and the squiggles and shapes leave you with a heavy, toxic feeling that poisons your confidence and undermines your sense of the world. English has never made less sense.

"That's an assumption," says the camper scientist when Bogacz finishes.

"You can also get the same effect from mutual inhibition," offers Bogacz.

"Inhibition is equivalent to the activation of that neuron?"

"No, no," says Bogacz. "This is the *Shadlen* model." Bogacz then skips to another part of the poster. The camper scientist squeezes past me to get closer, and I am pushed out of the circle.

I go to the snack bar. The menu— simple, understandable—reads like poetry. After ordering a large Coca-Cola, I find a spot near Bogacz's poster and try to regroup. Is my tongue swelling? Can listening to something incomprehensible make your tongue swell? I am incredibly, crushingly tired, but I refuse to give up.

I look at my notebook and cross out *Attend as a peer*, then I head back to the poster as a regular old journalist. Surely if I listen carefully and apply my interviewing skills, then I will be able to understand, but after three run-throughs and a multitude of questions I still have no idea what this poster is about. There are monkeys and statistical models and this whole third part that involves some kind of video game that college students played for pretend money.

"Is it that the eye is 'deciding'?" I say during the third run-through. "But it's not a conscious decision?"

I just want to understand *something*, but Bogacz isn't sure how to answer. ▶

66 Is my tongue swelling? Can listening to something incomprehensible make your tongue swell? I am incredibly, crushingly tired, but I refuse to give up. 99

Head Case *(continued)*

"We're conscious of what we are saying, but are we conscious of where the words are coming from?" says a friendly brain scientist who is standing next to me.

For a moment I think I have it . . . yes, there it is . . . but then it slips away. By now I've been here an hour—the entire poster session will be over before lunch—and I have a penetrating headache that will stay with me for the rest of the day. Bogacz starts over from the beginning for someone else, and I tell myself I have to go, I have other posters to visit but I can't leave. Bogacz has drawn an even larger crowd, and that must mean something. There is meaning here. I just don't see it yet.

Since science journalism is failing me, I am left with only one other option: snark. I start taking notes on Bogacz's five o'clock shadow and the garbage at the foot of every poster stand, and how the neuroscientist at the poster next door is getting no play at all and has to stand there like a wallflower at a school dance.

Eventually I abandon this approach, and I make what I hope is an optimal decision: Time to move on. As I'm walking away I run into the friendly brain scientist who offered up the koan about being aware of where our words come from. Given how many disputes Bogacz was fending off, I have one final, final, final way to save face: Rip on Bogacz's poster and hope that the Zen master agrees.

66 Since science journalism is failing me, I am left with only one other option: snark. 99

14

"Oh, no," he says, like I'm the one who is crazy. "This is good stuff."

With that I'm done.

Thomas Hardy once noted that "what we gain by science is, after all, sadness." He was speaking about science's power to demythologize the universe, how "ghastly a business we perceive it all to be—and the non-necessity of it." But science doesn't have to do anything so grandiose to bring a person down. I didn't know it at the time, but that one poster would do me in for the whole day. I went on to my next poster and the next and repeated this experience to some degree or another ten more times that day. That evening I staggered to my hotel room and collapsed on the bed and passed out cold in my clothes at 7:00 p.m.

From trying. ∽

Three Three-Sentence Reviews of Books and Articles About the Brain by People Who Are Far, Far Smarter Than Dennis Cass

THE FEELING OF WHAT HAPPENS: BODY AND EMOTION IN THE MAKING OF CONSCIOUSNESS by Antonio Damasio

Scientifically speaking, emotions and feelings are related but different. How they are related but different is both fascinating in the abstract and useful in everyday life. Damasio can be a little pretentious, but for most part he delivers the goods.

THE BLANK SLATE: THE MODERN DENIAL OF HUMAN NATURE by Steven Pinker

The popular wisdom says that people are born with minds that are blank slates. But that's not even remotely true. For five hundred pages Pinker tells you why and does it in a way that is both charming and impressive.

"EVOLUTIONARY PSYCHOLOGY: A PRIMER" by Leda Cosmides and John Tooby

You can find this article on the Web at http://cogweb.ucla.edu/ep/EP-primer .html. Print it up, pour yourself a large soda, and then sit down and learn why our "modern skulls house a Stone Age mind." Contains numerous concepts with which to astound friends and delight dinner companions.

> **"** The popular wisdom says that people are born with minds that are blank slates. But that's not even remotely true. **"**